A Practical Introduction to Index Numbers

A Practical Introduction to Index Numbers

Jeff Ralph

Office for National Statistics, UK

Rob O'Neill

University of Huddersfield, UK

Joe Winton

Office for National Statistics, UK

Library of Congress Cataloging-in-Publication Data

Ralph, Jeff.
   A practical introduction to index numbers / Jeff Ralph, Rob O'Neill, Joe Winton.
      pages cm
   Includes bibliographical references and index.
   ISBN 978-1-118-97781-1 (paperback)
   1. Index numbers (Economics)   I. O'Neill, Rob.   II. Winton, Joe.   III. Title.
   HB225.R35 2015
   330.01′5195–dc23
                                                                    2015019379

A catalogue record for this book is available from the British Library.

Typeset in 10/12pt TimesLTStd by SPi Global, Chennai, India

Printed in Singapore by C.O.S. Printers Pte Ltd

1   2015

Contents

Preface

The inspiration for writing an introduction to index numbers arose from our work in delivering index number training. We provide introductory and advanced training courses for the Government Statistical Service, for the Official Statistics module of the Operational Research, Applied Statistics and Risk MSc course at Cardiff University and for students sitting the index number components of the Royal Statistical Society (RSS) Ordinary and Higher Certificates. A number of students have asked us to recommend an introductory text to supplement the training, and we have struggled to find a suitable modern book to recommend. While other statistical topics are well supported by introductory books, this is not the case for index numbers.

There are, of course, a number of very good books on index numbers; however, they are directed towards more advanced study. We wanted to write a short, practical book that would introduce the subject and be suitable for students and others who would like a general introduction to the subject. We have therefore based the content partly on the syllabuses for the two RSS Certificates and partly on the content of the workshops we have developed at the Office for National Statistics.

Index numbers is a topic whose concepts are applied very widely. In the economic sphere, many of the most prominent of our economic indicators are presented as index numbers, including the Consumer Prices Index, Gross Domestic Product and the Retail Sales Index. Readers of most newspapers will find regular mentions of these indices and many others too. However, it is not just in the economic sphere that index number concepts appear. Increasingly, they are finding application in the social sphere through composite measures such as those aiming to quantify complex concepts such as poverty and prosperity.

Despite finding widespread application, few statistics or economics books contain even a mention of index numbers, let alone any detailed description; a curious student might well wonder which academic subject they fall under. The answer is that the topic tends to fall between economics and statistics. Where it does appear, it is usually studied as part of a specialist topic, like measurement economics.

At a basic level, the subject may appear simple to the beginning student, or casual academic observer; however, in our experience this is not the case. There are many subtle and complex aspects that have gradually developed over many years and are still the subject of much discussion and academic research.[1] Indeed, the practical implementation of a price index like the Consumer Prices Index is a massive undertaking that requires a considerable amount of effort and expertise on the part of National Statistics Institutes around the world. It is the intention of this book to highlight some of these aspects, hopefully resulting in an enhanced understanding of index numbers as commonly used in Official Statistics.

In common with other topics in statistics and economics, some observers may consider index numbers a dry subject, distant from the general population and perhaps not very interesting. While we can understand why this view might be held, we see the subject very differently. It is a complex and absorbing subject that provides a good return on investment in study. It should also be pointed out that through index number measures such as the Consumer Prices Index, all of our lives are affected in one way or another by the application of the subject.

This is not a book that will examine and explain all of the complex aspects of major index number outputs. However, there are facets of the Consumer Prices Index that are instructive for the student to consider, and we have included some of them in this book. We hope they will also prove interesting and will give the student a better understanding of this key indicator when watching the news, listening to the radio or reading a newspaper.

One of the significant advantages of developing and delivering training courses over a number of years is the opportunity to continuously improve the teaching material based on feedback from students. Two particular aspects of the training have proved popular with students. Firstly, including exercises that require students to calculate index numbers from a modest amount of data, and secondly, mixing material on the theoretical aspects of the subject with sections on the practical aspects of price statistics. We have continued this approach in this book.

Organisation of the book

This book has 15 chapters and includes 11 sets of exercises. Chapter 1 provides a general introduction to the concepts of an index and an index number; it identifies a range of applications from both the economic and social spheres. Chapter 2 introduces the reader to the procedure for converting a simple data series into a series of index numbers and back again. It shows how to calculate percentage change from an index series and identifies some benefits of working with data in index number form. The first exercise gives the reader the opportunity to develop the basic skills with some simple data.

One of the major applications of index numbers is the measurement of inflation; it is arguably the most important output any National Statistics Institute produces and it is the subject of Chapter 3. We explain briefly what is meant by inflation and why it is important. Chapters 4 and 5 set out the basic material to enable the student to understand how inflation is calculated in practice. Chapter 4 introduces

---

[1] For example, see the Johnson Review of UK Consumer Price Statistics: http://www.statisticsauthority.gov.uk/reports---correspondence/current-reviews/range-of-prices-statistics.html.

price, quantity and value, and the application of simple unweighted price indices. An exercise follows in which the reader is invited to calculate price index numbers for goods on sale in a shop. Chapter 5 describes value change for a basket of goods between two time periods and introduces the iconic price (and quantity) indices of Étienne Laspeyres and Hermann Paasche. It explains the difference between the two indices and shows how value change can be decomposed into measures of pure price change and pure quantity change; it also introduces the index number problem. An exercise follows.

Chapters 6 and 7 explore two aspects of the subject that are important for the practical application of index numbers – domains and aggregation, and linking and chain-linking. How the Consumer Prices Index is constructed is the subject of Chapter 8; it uses the knowledge gained from Chapters 2 to 7. Two more practical topics are covered in Chapters 9 and 10 – re-referencing and rebasing, and deflation; both chapters are followed by exercises. A brief description of the major price and quantity indices produced by National Statistics Institutes is given in Chapter 11.

Chapter 12 returns to the theoretical development of the subject, considering further index formulae including the Fisher, Walsh and Törnqvist indices; an exercise includes both algebra and calculation based questions. Chapter 13 gives an overview of the various approaches that have been made to identify the 'best' formula to use. More advanced topics that are much discussed and studied in the index number research field are the subject of Chapter 14. They include: consumer substitution behaviour, new goods and disappearing goods, and hard to measure goods like housing. These are all challenging topics that producers of price indices need to address and we seek only to introduce them in this chapter. Current research topics, such as the role of big data in price indices are explored in Chapter 15.

Six appendices provide supporting material:

- **A: Mathematics for index numbers**
  Index numbers is a mathematical topic – though the level of mathematical knowledge needed is relatively modest at the introductory level that this book covers. However, we do not want readers to be inhibited by the mathematical requirements, so we have explained the mathematical notation and knowledge required in each chapter as it arises. Appendix A provides further support by providing a fuller explanation of the key mathematical concepts and index number results.

- **B: Choice of index formula**
  This appendix provides a more detailed look at two of the approaches to index numbers that were introduced in Chapter 13 – the axiomatic and economic. This appendix is more advanced than any of the other material in the book.

- **C: Glossary of terms and formulae**

- **D: Solutions to most of the exercises**

- **E: Further reading**

Additional material available online

The companion website, www.wiley.com/go/ralph/index_numbers, hosts additional content for this book. This includes presentations, example data and R code to run index number calculations. There is also some additional help for students who attempt the exercises.

Suggested routes through the book

While it is hoped that many readers will work through all chapters in the book, there are options for students taking the Royal Statistical Society exams[2] who want to focus on the specific material they need. The following chapters are particularly relevant:

Ordinary certificate: 1–5 and some parts of 7 and 12

Higher certificate: 1–13 and Appendix A.

For a reader who would like to learn about the concept of index numbers, their use and some of the challenging aspects, but without wanting to develop skills in calculation, the following is a suggested subset of the book: Chapters 1–3, 8, 11, 14 and 15.

---

[2] More information on these exams and the qualifications they relate to can be found at https://www. rss.org.uk

Acknowledgements

We would like to thank our colleagues (and former colleagues) at the Office for National Statistics, in particular, those who work in the Methodology and Prices Divisions. Over the years, we have had many helpful conversations with them about both theoretical and practical aspects of index numbers. Thanks also go to the many students who have attended the courses and workshops we have given for the Government Statistical Service and at Cardiff University; their feedback has also been of great benefit to us.

A small army of kind colleagues read and commented on chapters of the book. From outside the Office for National Statistics, we would like to thank Emeritus Professor Bert Balk of Erasmus University, Professor Caroline Elliott of Huddersfield University and Associate Professor Paul Smith of Southampton University. From within the Office for National Statistics, we would like to thank Duncan Elliott, Dr Ria Sanderson, Dr Gareth Clews, Jim O'Donoghue, Derek Bird, Richard Campbell and Ainslie Restieaux. Thanks also go to Laura Clarke and Lauren Archer who read an early complete draft as their introduction to the subject. In the time-honoured manner, any remaining errors in the book are entirely the responsibility of the authors.

We would also like to thank our partners Bryony, Sarah and Becky for their patience as well as their invaluable comments and advice throughout our writing. Thanks also go to Wiley publishers for their expert handling of the manuscript.

Finally, we would like to note that the views expressed in this book are those of the authors and not necessarily those of the organisations for which they work.

1

Introduction

1.1 What is an index number?

The simplest description of an index number is that it is a measure of change. Consider the data in Table 1.1, which shows the total value of retail sales[1] for Great Britain between 2005 and 2008 presented in two ways, firstly, as values in billion pounds, and secondly, scaled so that the value in 2005 is set to be 100.

The idea behind representing the time series of the values of sales in a scaled form is to make the degree of change readily apparent. The process of creating values in the third column is a simple one. Firstly, we choose a time period as the reference time period with which we want to compare the change; in this case, we have chosen 2005 as the reference (or base) time period. The index number series is then scaled to be equal to 100 for this reference period; the same scaling factor is then applied to the values of sales for other years. We explain how to do this in detail in Chapter 2.

The values for the scaled series are set to be around 100, as this is judged to make the degree of change clearest; the scaled values are called index numbers. Representing the time series in this way makes comparison easy. For example, the percentage change in retail sales between 2007 and 2005 can just be read from the index number for 2007 – it is 7.88%. Note that although the scaling process changes the numbers, it does not alter the percentage differences. Chapter 2 shows how to convert the percentage change from an index series back to values; for example, if we want to calculate how much money the change of 7.88% in this series represents.

By creating an index number representation of the time series of retail sales values, we have gained a more direct representation of change. In doing so, we have lost the actual monetary values; however, frequently the focus is primarily on the change in the level of the series rather than on the actual amount sold in billion pounds.

---

[1] As the name suggests, this is the total estimated value of sales across retailers in Great Britain in any given year. More information on this series can be found at http://www.ons.gov.uk/ons/rel/rsi/retail -sales/index.html

---

A Practical Introduction to Index Numbers, First Edition. Jeff Ralph, Rob O'Neill and Joe Winton.
© 2015 John Wiley & Sons, Ltd. Published 2015 by John Wiley & Sons, Ltd.
Companion Website: http://www.wiley.com/go/ralph/index_numbers

Table 1.1   Value of retail sales 2005–2008 for Great Britain.

	Value of retail sales (£bn)	Value relative to sales in 2005 (2005 = 100)
2005	281.450	100.00
2006	292.110	103.79
2007	303.621	107.88
2008	321.178	114.12

Source: Office for National Statistics (Time series of retail sales data are available from the ONS website http://www.ons.gov.uk/ons/rel/rsi/ retail-sales/july-2013/rft-rsi-poundsdata-july-2013.xls; series ValNSAT).

1.2   Example – the Consumer Prices Index

A different example of an index number series is provided by the Consumer Prices Index (CPI). This is a measure that tracks the movement in the general level of prices of consumer goods and services.

Table 1.2, taken from the CPI Statistical Bulletin for September 2013, shows index numbers representing the general level of prices for each month from September 2012 to September 2013,[2] where the index value has been set to be 100 in 2005. The index number represents the general level of prices in any given month. The change between the level of prices in any given month and the level of prices in 2005 is easily found by referring to the index number. For example, in September 2012, we can see that prices had increased by 23.5% from 2005.

Table 1.2 also contains the percentage rate of change in the general price level for each month compared with the previous month ('1-month rate') and compared with the same month in the previous year ('12-month rate'). The two rates of change figures are very important as the rate of change of the general price level is also known as the rate of inflation.

The monthly figure for the 12-month rate is the headline inflation figure produced by the UK Office for National Statistics and is arguably the most important of all economic statistics. When it is released each month, it is often the lead item in news bulletins and is reported widely in the national press. Its prominence is a consequence of its widespread use as a key input to important economic decisions such as setting interest rates and its use in adjusting benefits and allowances.

Producing the CPI every month is a considerable task, which is the responsibility of the Prices Division within the UK Office for National Statistics. Ideally, every transaction for every good and service in the UK would be captured over the month and an average of the price paid across all these transactions would be taken. This is, however, clearly not possible, and hence a sample of prices of goods and services is taken instead. The UK Office for National Statistics constructs a representative

---

[2] See http://www.ons.gov.uk/ons/rel/cpi/consumer-price-indices/september-2013/stb---consumer-price -indices---september-2013.html

Table 1.2   CPI values, 1- and 12-month inflation rates: September 2012–2013, United Kingdom.

		Index (UK, 2005 = 100)	1-Month rate	12-Month rate
2012	Sep	123.5	0.4	2.2
	Oct	124.2	0.5	2.7
	Nov	124.4	0.2	2.7
	Dec	125.0	0.5	2.7
2013	Jan	124.4	−0.5	2.7
	Feb	125.2	0.7	2.8
	Mar	125.6	0.3	2.8
	Apr	125.9	0.2	2.4
	May	126.1	0.2	2.7
	Jun	125.9	−0.2	2.9
	Jul	125.8	0.0	2.8
	Aug	126.4	0.4	2.7
	Sep	126.8	0.4	2.7

Source: Office for National Statistics.

'basket' of goods and services and records the prices of these items from a sample of geographical locations and from a sample of shops within these locations. The sample is constructed to represent the range of shops and locations in a 'fair' way. The basket consists of about 700 representative goods and services and about 180 000 prices of these items are captured every month. More information on how the CPI is calculated is given in Chapter 8.

It is interesting to note here that the value of the general level of prices is not a particularly useful statistic and is not published; it is the rate of change that is the statistic which is important. Of course, we could look at the price level of the basket and how it changes over time; however, it contains such a variety of goods and services that it would be difficult to interpret.

Regarding the CPI, it is not just the headline figure of the rate of inflation for the UK for all goods and services that is of interest. The inflation figure is made up of changes in the costs of different types of goods and services and their price movements are different. The following graph shows the index numbers for four types of goods and services for a 2-year period, where the index numbers have been scaled to be 100 in September 2011.

Figure 1.1 shows that the variations in prices of different types of goods and services are different. For example, there has been a greater rise in prices for 'Alcoholic Beverages and Tobacco' than for 'Transport' for most time periods. The solid black line shows the all-items index series, which is a combination of index series covering 12 types of goods and services (only four types are shown in Figure 1.1), and so the movement of the overall index series is an 'average' of the variations in the 12 index series for more specific categories of goods and services. Note that it is not just the

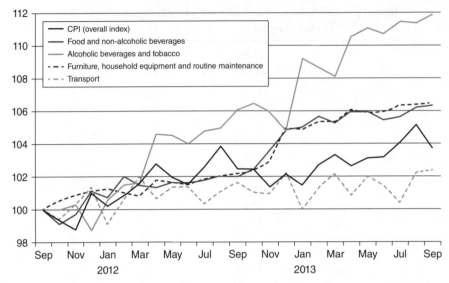

Figure 1.1 UK Consumer Prices Index and selected sectors, September 2011–2013. Source: Office for National Statistics (http://www.ons.gov.uk/ons/rel/rsi /retail-sales/july-2013/tsd-retail-sales--july-2013.html).

overall, or the 'all-items' CPI index number series, that is useful; each of the component index number series is also valuable. In addition, depicting the component series as index numbers makes comparison of their movements clearer; if the figure had shown the price levels, they would be much more difficult to compare.

The index numbers in the all-items CPI are not a simple average of the index numbers of the 12 sub-categories; they are combined using the proportion of consumer expenditure on each type of commodity to give a 'weight' to each sub-category. Table 1.3 shows the 12 categories and the weights assigned for the year 2013; these weights are revised every year in line with other information collected by the UK Office for National Statistics, which looks at the spending patterns of UK households. Weights are applied so that the variations in the component series are combined in a 'fair' way to produce an overall variation in the level of prices.

Having seen some examples, we are now ready to consider a definition of an index number. One of the most famous definitions was given by the economist Irving Fisher (1867–1947), who was highly influential in the development of index numbers as a subject and in economics in general. He established his own index number institute as a business that sold economic data in index number form, long before governments became the main suppliers of index numbers [1]. He stated that:

> The fundamental purpose of an index number is that it shall fairly represent, so far as one figure can, the general trend of the many diverging ratios of which it is composed.
>
> (Fisher [2])

Table 1.3   Weights for categories of goods and services in the CPI 2013, UK.

	Parts per 1000
01 Food and non-alcoholic beverages	95
02 Alcoholic beverages and tobacco	38
03 Clothing and footwear	60
04 Housing, water, electricity, gas and other fuels	244
05 Furniture, household equipment and maintenance	52
06 Health	23
07 Transport	128
08 Communication	26
09 Recreation and culture	123
10 Education	18
11 Restaurants and hotels	103
12 Miscellaneous goods and services	90
	1000

Source: Office for National Statistics (www.ons.gov.uk/guide-method/user-guidance/prices/cpi-and-rpi--updating-weights/2013.pdf).

How well does this definition apply to the CPI example? Firstly, the CPI price index numbers summarise the differing movements of the prices of constituent goods and services into one overall index series. Secondly, it is a fair measure in the sense that the price movements are weighted by the expenditure shares of each type of good and service.

1.3   Example – FTSE 100

The FTSE 100[3] is one of a wide range of stock market indices produced by the FTSE Group. It is an index of the share prices of the top 100 UK companies as measured by market capitalisation. It was first produced in 1984 when the index value was set to 1000 and since then, it has not been re-scaled; its current value is more than 6700.[4] The index is updated every 15 seconds during trading hours. The share price for each company is weighted by a factor involving the size of the company as measured by the market capitalisation (the total value of shares).

How does this index fit in with Irving Fisher's definition? The index certainly summarises a great deal of share information, where the movement of each share is different. It is fair in the sense that each individual share price movement is weighted by a factor that includes the market capitalisation of the company. The index was not initially set to 100; instead it was set to 1000. However, there is no strict need to take an initial value of 100; it is just a convenient value. In a similar way, the index value has reached above 6000, a significantly larger number than the initial value.

---

[3] See   http://www.londonstockexchange.com/exchange/prices-and-markets/stocks/indices/summary/summary-indices.html?index=UKX
[4] At the end of July 2014.

Table 1.4   Multidimensional Poverty Index dimensions, indicators and weights.

Dimension	Indicator	Weight
Health	Child mortality	1/6
	Health	1/6
Education	Years of schooling	1/6
	Child school attendance	1/6
Standard of living	Electricity	1/18
	Drinking water	1/18
	Sanitation	1/18
	Flooring	1/18
	Cooking fuels	1/18
	Assets	1/18

1.4   Example – Multidimensional Poverty Index

For an indicator such as inflation, we are combining together quantities of the same type, which are price changes for goods and services. This process is called aggregation and provides an overall, summary measure of the change in the general level of prices. This index number concept of summarising changes has been applied to wider uses where a single summary measure is created from different quantities having different units to give what is known as a 'composite indicator'.

To illustrate the concept of an index number as a composite, we consider a measure of poverty. The Multidimensional Poverty Index is a measure of acute poverty produced by the Oxford Poverty and Human Development Initiative (OPHI, Oxford [3]). It combines a number of deprivations that an individual may experience at any one time to provide a comprehensive measure of extreme deprivation.

The measure is composed of dimensions, indicators and weights, which are summarised in Table 1.4.

To produce a numerical score, criteria are applied to each indicator; the resulting scores are then combined into an overall poverty measure.[5] This facilitates the comparison of poverty scores across countries and regions and also across time.

1.5   Example – Gender Inequality Index

A second example of an index number, as a composite measure, is provided by the Gender Inequality Index, produced by the United Nations as part of the United Nations Development Programme.[6] It is a measure of the inequality in achievement by men and women and is calculated every year; in 2013, it covered 187 countries.

---

[5] See http://www.ophi.org.uk/

[6] United Nations Development Programme, Gender Inequality Index, http://hdr.undp.org/en/content/gender-inequality-index-gii

Table 1.5   Gender Inequality Index, dimensions and indicators.

Dimension	Indicator
Health	Maternal mortality ratio
	Adolescent birth rate
Empowerment	Parliamentary representation
	Attainment at secondary and higher education
Labour market	Labour market participation

Of course, 'gender inequality' is not a quantity that can be measured directly. A numerical score for this concept is constructed by specifying and measuring attributes of gender inequality, which experts consider to be relevant and which have a resonance with the general public. The Gender Inequality Index has three dimensions and five indicators (Table 1.5).

For each country, a numerical value is assigned to each indicator and the values are combined into a single overall score.[7] The scores for each country can be ranked to identify which countries display the most and which display the least gender inequality. The figures for 2013 show that European countries tend to display small values of gender inequality while African countries mostly show the highest values.

1.6   Representing the world with index numbers

Although the methodologies behind all aggregates and composite indicators are disputed to a degree, those in the social, political and medical domains attract greater attention to the judgements made in order to construct index numbers. As a result, commentators have expressed concern about their validity and have suggested that we should treat them with a degree of scepticism. As the number of indicators has increased along with the experience of using them, there has been much academic work to identify the best practice in the design and construction of such indicators [4]. For example, it is important that the theoretical background to a measure is based on solid scientific research and evidence; this will result in a robust mathematical model to link the aspect of the world being examined to the factors that influence it. In addition, the data used in the indicators need to be measured in an accurate manner.

The construction of indicators, whether social, medical, political or economic, follows the same overall structure: an indicator is constructed from raw or processed data combined in accordance with a model of the aspect of the world being considered. It might be thought that a measure such as CPI does not require a theoretical framework; however, as Chapter 8 will show, there is much discussion over the theoretical basis of a measure of inflation.

---

[7] United Nations Development Programme, Table 1.4. http://hdr.undp.org/en/content/table-4-gender-inequality-index

For composite indicators constructed to measure subtle, abstract quantities, there are both positive and negative aspects. On the positive side, they can summarise complex phenomena in a single index number to support communication and policy making. With a time series of such measures, an assessment of the progress (or the lack of it) can be made. On the negative side, they can lead to a simplistic representation of complex phenomena, which can then lead to inappropriate policy decisions. A poor model and/or poor data will result in a poor indicator. The weighing of different components is a particularly difficult topic.

1.7   Chapter summary

Although many important phenomena are difficult to measure, the advantages of a numerical representation are significant. Policy makers frequently want to allocate limited resources to achieve the highest impact and they want a means to assess the impact of policy decisions. This need for quantification of an ever wider range of aspects of the world has led to an increasing number of indicators being devised and this trend will surely only continue. An optimistic view of the future would suggest that as our experience with using such indicators grows, so will our ability to develop better quality measures.

References

1. See Vogt, A., Fisher, I., and Barta, J. (1997) *The Making of Tests for Index Numbers: Mathematical Methods of Descriptive Statistics: Published in Honour of the 50th Anniversary of the Death of Irving Fisher*, Physica-Verlag, Heidelberg.
2. Quote from Fisher, I. (1922) *The Making of Index Numbers: A Study of their Varieties, Tests, and Reliability*, Mifflin, Boston, MA. Californian Digital Library, https://archive.org/details/makingofindexnum00fishrich (accessed 16 January 2015), p. 10.
3. See University of Oxford (2014) Oxford Poverty and Human Development Initiative, http://www.ophi.org.uk/research/multidimensional-poverty/ (accessed 13 January 2014).
4. See OECD, JRC, European Commission, OECD, Joint Research Centre, European Union, and Source OECD (Online service) (2008) *Handbook on Constructing Composite Indicators: Methodology and User Guide*, Organisation for Economic Cooperation and Development (OECD), Paris. http://www.oecd.org/std/42495745.pdf

2

Index numbers and change

Chapter 1 introduced the general concept of an index number as a measure that can combine similar or different types of data together in a fair way leading to a single, summary value that facilitates clearer comparison between time periods. More formally, an index number is a measure that shows change in a variable or group of variables with respect to a characteristic such as time. A collection of index numbers across characteristics (e.g. time) is called an index series.

This chapter shows how to create an index number series from a data series and vice versa. It also shows how to calculate percentage changes from an index series.

2.1 Calculating an index series from a data series

The most basic form of an index number is a simple relative index series; in this chapter, we consider only this simplified index number form in order to focus on the technique of converting to and from index numbers.

To convert a time series (or any other data series) to an index series, we first choose one period as the base period (usually labelled 'period 0') and then for every other observation of the variable, we divide the value of the series by the value in the base period. For ease of comparison, index numbers are often expressed with the value in the base period set to 100; to do this, the value of the index is multiplied by 100 – this is optional but common.

Let x^t represent the value of a variable x at a time period t and x^0 represent the value of the variable x at time period 0. Then, the simple relative index for the variable x at the current time period t with base period 0 is given by:

$$I^{0,t} = 100 \times \frac{x^t}{x^0} \tag{2.1}$$

From this equation, it can be seen that the index with time period t set to be the base period 0 will have a value of 100 and if the variable x increases in later time periods, the value of the index will be greater than 100.

A Practical Introduction to Index Numbers, First Edition. Jeff Ralph, Rob O'Neill and Joe Winton.
© 2015 John Wiley & Sons, Ltd. Published 2015 by John Wiley & Sons, Ltd.
Companion Website: http://www.wiley.com/go/ralph/index_numbers

Table 2.1   Annual turnover for Company A, 2002–2013.

						£ million
Year	2002	2003	2004	2005	2006	2007
Turnover	17.9	18.5	18.1	18.6	19.0	19.1
Year	2008	2009	2010	2011	2012	2013
Turnover	19.3	20.2	20.3	20.0	20.5	20.4

Example 2.1 shows how to use this formula to calculate an index series from a data series.

Example 2.1   Table 2.1 shows the annual turnover for a company: Company A. Calculate a simple relative index series of turnover for Company A, with 2002 as the base period.

Solution For each period, we need to calculate a simple relative index with 2002 as the base period (period 0):

$$I^{0,t} = 100 \times \frac{x^t}{x^0}$$

So for $t = 2002$:

$$I^{0,t} = I^{2002,2002} = 100 \times \frac{x^{2002}}{x^{2002}} = 100 \times \frac{17.9}{17.9} = 100$$

and for 2003:

$$I^{0,t} = I^{2002,2003} = 100 \times \frac{x^{2003}}{x^{2002}} = 100 \times \frac{18.5}{17.9} = 103.5 \ (1\,\text{d.p.})$$

and for 2012:

$$I^{0,t} = I^{2002,2012} = 100 \times \frac{x^{2012}}{x^{2002}} = 100 \times \frac{20.5}{17.9} = 114.5 \ (1\,\text{d.p.})$$

And the rest of the series:

Year	Turnover	Index number
2002	17.9	$100 \times (17.9/17.9) = 100.0$
2003	18.5	$100 \times (18.5/17.9) = 103.4$
2004	18.1	$100 \times (18.1/17.9) = 101.1$
2005	18.6	$100 \times (18.6/17.9) = 103.9$
2006	19.0	$100 \times (19.0/17.9) = 106.1$
2007	19.1	$100 \times (19.1/17.9) = 106.7$
2008	19.3	$100 \times (19.3/17.9) = 107.8$

(*continued overleaf*)

Year	Turnover	Index number
2009	20.2	$100 \times (20.2/17.9) = 112.8$
2010	20.3	$100 \times (20.3/17.9) = 113.4$
2011	20.0	$100 \times (20.0/17.9) = 111.7$
2012	20.5	$100 \times (20.5/17.9) = 114.5$
2013	20.4	$100 \times (20.4/17.9) = 114.0$

2.2   Calculating percentage change

Index numbers are measures of change. In an index series, measurements are expressed as the change in the series from a fixed point (or base) to the current point. When interpreting index numbers, it is often not the value (or the level) of the index that we are interested in, but rather the change (or the growth) in that index series. In fact, in the UK, the headline figures reported by the Office for National Statistics for both Gross Domestic Product (GDP) and the Consumer Prices Index (CPI) are changes in the index series.[1] Such changes are usually expressed as a percentage.

The percentage change in an index series is not only easier to understand than the level of an index series but also in the case of the outputs mentioned above provides a clearer snapshot of how the economy is performing. It is, therefore, important that we can calculate the percentage change in a series.

The percentage change in a series between point A and point B is calculated as:

$$\text{percentage change} = \frac{B - A}{A} \times 100 \tag{2.2}$$

That is, the difference between the levels of the index series at the two time points, divided by the value of the series at the earlier time point, multiplied by 100 in order to express the value as a percentage.

In the case of index numbers, we can calculate the growth in an index between two periods, time s and time t (where $s < t$) as:

$$g^{s,t} = \frac{(I^{0,t} - I^{0,s})}{I^{0,s}} \times 100 \tag{2.3}$$

When calculating the percentage change in an index series from the base period 0 to any other period t in the series, this calculation is straightforward[2] and highlights one of the reasons for choosing 100 as the value of the index in the base period:

$$g^{0,t} = \frac{(I^{0,t} - I^{0,0})}{I^{0,0}} \times 100 = \frac{(I^{0,t} - 100)}{100} = I^{0,t} - 100$$

See Example 2.2 for a worked example of a percentage change calculation.

---

[1] This is also true for many other key social and economic indicators.

[2] This calculation is dependent on index numbers being expressed as a percentage. If $I^{0,0} = 1$ then $g^{0,t} = I^{0,t} - 1$.

Although we can calculate the growth or change in a series from one point to any other, the two most common calculations of growth in a monthly index series are when $t - s = 1$ (change from the previous month) and when $t - s = 12$ (change from the same month a year ago).

Table 2.2   Index of annual turnover for Company A, 2002–2013.

						$2002 = 100$
Year	2002	2003	2004	2005	2006	2007
Index	100.0	103.4	101.1	103.9	106.1	106.7
Year	2008	2009	2010	2011	2012	2013
Index	107.8	112.8	113.4	111.7	114.5	114.0

Example 2.2   Table 2.2 shows the index of turnover for Company A, with 2002 as the base period as calculated in Example 2.1.

Calculate the percentage change in the index between the following years:

a) 2002 and 2004

b) 2002 and 2013

c) 2005 and 2007

Solution Use: $g^{s,t} = \frac{(I^{0,t} - I^{0,s})}{I^{0,s}} \times 100$

a)
$$g^{2002,2004} = \frac{(I^{2002,2004} - I^{2002,2002})}{I^{2002,2002}} \times 100 = \frac{(101.1 - 100)}{100} \times 100$$
$$= 1.1 \ (1\,\text{d.p.})$$

b)
$$g^{2002,2013} = \frac{(I^{2002,2013} - I^{2002,2002})}{I^{2002,2002}} \times 100 = \frac{(114.0 - 100)}{100} \times 100$$
$$= 14.0 \ (1\,\text{d.p.})$$

c)
$$g^{2005,2007} = \frac{(I^{2002,2007} - I^{2002,2005})}{I^{2002,2005}} \times 100 = \frac{(106.7 - 103.9)}{103.9} \times 100$$
$$= 2.7 \ (1\,\text{d.p.})$$

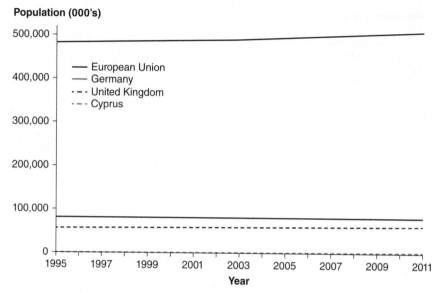

Figure 2.1   Population on 1st January, European Union, 1995–2011.   Source: Eurostat. These graphs have been created using data from the Eurostat Statistics Database: http://epp.eurostat.ec.europa.eu/portal/page/portal/statistics/search _database. Eurostat is the statistical office of the European Union and it is a Web site that provides access to all Eurostat's statistical databases and the associated electronic publications.

2.3   Comparing data series with index numbers

Index numbers can be used to make effective comparisons between different series.

It is often the case that data series have quite different values, which can make comparison difficult. By creating index number versions with a common base period, comparisons can be made more easily.

The chart in Figure 2.1 shows the total population of the European Union (EU) along with the populations of three EU countries across a 16-year period, 1995–2011. It is clear from Figure 2.1 that the population of the EU is much larger than those of Germany and the UK and the population of Cyprus is very small in relation to these larger countries.

Suppose we want to compare changes in the four populations over this time period. If we try to use the raw population figures, then this becomes difficult to see the relative changes. By converting the four series in Figure 2.1 to index number series and choosing a common base period, we can see the percentage changes in population across the 16-year period far more easily.

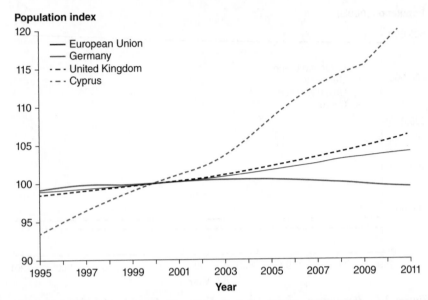

Figure 2.2   Index of population on 1 January (2000 = 100), European Union, 1995–2011.   Source: Eurostat. These graphs have been created using data from the Eurostat Statistics Database: http://epp.eurostat.ec.europa.eu/portal/page/portal /statistics/search_database. Eurostat is the statistical office of the European Union and it is a Web site that provides access to all Eurostat's statistical databases and the associated electronic publications.

From the chart in Figure 2.2, we can see that while the populations of the EU and the United Kingdom have risen steadily throughout the time period, the population of Germany has remained fairly stable and the population of Cyprus has risen dramatically, with a large surge in growth between 2003 and 2006. In fact, we can see that the population of Cyprus has risen in percentage terms more than three times the population of the United Kingdom – a fact that is completely obscured by the different measurement scales in Figure 2.1. Hence, by converting the series of population measures to an index series, we have facilitated a comparison between these areas, which reveals trends that we might have otherwise not been aware of

2.4   Converting from an index series to a data series

We have seen how to create an index series from a data series and why such a series might be useful, but there will also be times when we have an index series and we would like to know the values of the raw data series from which it was created; for example some people like to know the GDP of the UK measured in billions of pounds. To do this, we also need to know the value of the data series at one time period for which we have the index series.

The value of a data series from which an index number was calculated at period t (x^t) is calculated as:

$$x^t = \frac{I^{0,t} \times x^s}{I^{0,s}} \qquad (2.4)$$

where x^s and x^t are values of the data series at some periods s and t and the $I^{0,s}$ and $I^{0,t}$ are the values of the index series at these two time points.

This means that in period t, the value of the data series from which an index number was calculated is found by multiplying the value of the index series in period t by a known value of the data series (at some period s), divided by the value of the index series in the same period.

Often, the known value of the data series is the value in the base period (x^0) which simplifies the above formula to:

$$x^t = \frac{I^{0,t} \times x^0}{100} \qquad (2.5)$$

Table 2.3 Index of annual turnover for Company A, 2002–2013.

						2002 = 100
Year	2002	2003	2004	2005	2006	2007
Index	100.0	103.4	101.1	103.9	106.1	106.7
Year	2008	2009	2010	2011	2012	2013
Index	107.8	112.8	113.4	111.7	114.5	114.0

Example 2.3 Table 2.3 shows the index of turnover for Company A, with 2002 as the base period as calculated in Example 2.1.

The turnover for Company A in 2006 was £19.0 million. Using just this information, calculate the turnover for the remaining years, 2002–2013.

Solution For each period, we need to find the value of the original data series (turnover) using the index series and the value of the data series at one period:

$$x^t = \frac{I^{0,t} \times x^s}{I^{0,s}}$$

We know that the turnover for Company A in 2006 was £19.0 million, so we can use $x^{2006} = 19.0$. So for 2002:

$$x^t = x^{2002} = \frac{I^{2002,2002} \times x^{2006}}{I^{2002,2006}} = \frac{100 \times 19.0}{106.1} = 17.9$$

and for 2003:

$$x^t = x^{2003} = \frac{I^{2002,2003} \times x^{2006}}{I^{2002,2006}} = \frac{103.4 \times 19.0}{106.1} = 18.5$$

And the rest of the series:

Year	Index number	Turnover (£ million)
2002	100.0	$(100.0 \times 19.0)/106.1 = 17.9$
2003	103.4	$(103.4 \times 19.0)/106.1 = 18.5$
2004	101.1	$(101.1 \times 19.0)/106.1 = 18.1$
2005	103.9	$(103.9 \times 19.0)/106.1 = 18.6$
2006	106.1	$(106.1 \times 19.0)/106.1 = 19.0$
2007	106.7	$(106.7 \times 19.0)/106.1 = 19.1$
2008	107.8	$(107.8 \times 19.0)/106.1 = 19.3$
2009	112.8	$(112.8 \times 19.0)/106.1 = 20.2$
2010	113.4	$(113.4 \times 19.0)/106.1 = 20.3$
2011	111.7	$(111.7 \times 19.0)/106.1 = 20.0$
2012	114.5	$(114.5 \times 19.0)/106.1 = 20.5$
2013	114.0	$(114.0 \times 19.0)/106.1 = 20.4$

2.5   Chapter summary

In this chapter, we have seen how to create an index number series from a data series; we have also seen how to calculate percentage growth and how to convert an index number series back into a data series. The chapter also explained how index numbers can help us compare measures that otherwise would not be easily comparable.

In later chapters, we shall consider the calculation of index numbers by involving many observations and some specific cases including price and quantity data.

Exercise A

Solutions to this exercise can be found in Appendix D.

A.1   Table A.1 presents the total number of employees for a particular company over the 10-year period, 2005–2014.

    a. For each year 2006–2014, calculate a simple index of employees, using 2005 as the base year.

    b. Calculate the percentage change in the index between 2005 and 2013.

    c. Calculate the percentage change in the index between 2008 and 2011.

Table A.1   Number of employees 2005–2014.

Year	2005	2006	2007	2008	2009
Total employees	192	192	204	224	220
Year	2010	2011	2012	2013	2014
Total employees	232	212	228	220	236

A.2   Table A.2 presents the total number of goals scored by all teams in a fictional football league each season from 2005/2006 to 2012/2013.

    a. Calculate a simple index of goals scored for each season with 2005/2006 as the base season.

    b. Calculate the percentage change in the index between 2005/2006 and 2012/2013.

    c. Calculate the percentage change in the index between 2006/2007 and 2010/2011.

    d. Calculate the percentage change in the index between 2009/2010 and 2012/2013.

Table A.2   Number of goals 2005/2006 to 2012/2013.

Season	2005/2006	2006/2007	2007/2008	2008/2009
Total number of goals	1215	1222	1217	1004
Season	2009/2010	2010/2011	2011/2012	2012/2013
Total number of goals	1001	1228	1215	1212

A.3   Richard has calculated a simple index of his salary over a 6-year period using 2009 as the base period; this series is presented in Table A.3.

Table A.3   Index of salary 2009–2014.

Year	2009	2010	2011	2012	2013	2014
Index of salary	100.0	112.0	112.0	120.0	128.0	132.0

Richard's colleague Catherine has a salary of £33 000 in 2014. Catherine would like to know what her salary was in 2009. Her salary has moved in exactly the same way as Richard's during this period.

a. Using the index of salary above, calculate Catherine's salary in 2009.

b. Calculate Catherine's salary in the remaining years, 2010–2013.

3

Measuring inflation

3.1 What is inflation?

According to the Palgrave dictionary of economics, inflation refers to a rise in the price level or a fall in the value of money [1]. It is one of the most often quoted and analysed macroeconomic statistics and is referred to in numerous contracts and agreements within the UK and between countries transacting in sterling. However, defining exactly what is meant by inflation is not easy. In this chapter, we discuss what we mean by inflation, why it is so important and thus emphasise why much of the discussion of index numbers presented in the rest of this book is centred on measuring changes in the price level.

The dictionary definition given above refers to a rising price level. Thus, inflation occurs when the price level, commonly referred to as the 'general price level', increases. Alternatively, it can be expressed as a fall in the value of money, or a fall in the purchasing power of a unit of currency. Deflation occurs when the general price level is falling and the value of money is rising.

Having defined inflation as a rise in the level of prices, or alternatively a fall in the value of money, the next obvious step is to question how we measure the general price level. An immediate question arises when we are considering a group of prices for very similar products, say the prices charged for exactly the same type of apple by two different shops; should we take an average of the two prices and if so how? The task becomes much more difficult as we include more and more of the products that people consume in a modern economy; for example how do we measure the price level of an economy that includes diverse products such as cars, computers and paperclips?

Inflation is an implicit variable, that is, it is never directly observed; however, it has very real consequences for everyone interacting in the real economy – that is, the part of the economy concerned with producing and consuming goods and services.

A Practical Introduction to Index Numbers, First Edition. Jeff Ralph, Rob O'Neill and Joe Winton.
© 2015 John Wiley & Sons, Ltd. Published 2015 by John Wiley & Sons, Ltd.
Companion Website: http://www.wiley.com/go/ralph/index_numbers

In the context of a UK consumer, inflation occurs when the value of the pound falls. As the economist Irving Fisher noted [2], there is a tendency for individuals to view a pound as a constant measure and prices as changing when in fact the process is a dual one. As prices rise, the pound in a person's pocket becomes worth less and is a store of ever decreasing value. It is this process that is most typically referred to when we talk about inflation.

When we consider how to measure inflation, we find that we have to make a number of important choices that are often taken for granted. For example, we need to specify both the group of people carrying out transactions and the type of economic transactions for which inflation is being measured. It is generally understood in most contexts that inflation refers to 'all (final) consumption transactions within a given national economic territory'. However, in an open economy, there will always be debate regarding which elements should be included in such a definition. For example, the main measure of inflation in the UK, the Consumer Prices Index (CPI), includes transactions made by consumers across the whole range of incomes. However, historically, the main measure was the Retail Prices Index (RPI), which does not include the transactions of those at the bottom and top of the income distribution.

When we use the term inflation, we assume that we are referring to something that is unambiguously defined. The truth is, however, that the practical, measured quantity incorporates a number of difficult choices that are still being debated. Even if we agree that we are trying to measure the change in the price level for a well-defined geographical area between well-defined time periods and with good data, there remain differences of technique that leads to small but important differences in the estimation of inflation, which can have a range of repercussions throughout the economy. It is intended that, by exploring some of these choices in this book, we can provide users of index numbers with a more thorough understanding of how changes in economic variables are measured.

3.2   What are inflation measures used for and why are they important?

Inflation measures are used in a wide range of contexts and any attempt at a complete definition of their uses would be near impossible; however, the list below provides an indication of some of the main uses of inflation measures and is intended to highlight the breadth of the applications to which statistics on the rising price level can be put:

- Determination of Monetary Policy by a Central Bank

- Adjustment to provisions for private pensions

- Adjustment of amounts paid over long-term contracts

- Adjustment of rail fares and other goods

- Evaluation of investment decisions

- Inputs to economic research and analysis

- Adjustment of index-linked debt

- Adjustment of tax allowances

- Targets for stability of the economy in an international context

Below, we shall try to give a flavour of each of these nine uses to which inflation measures are put. We shall describe how each requires an accurate measure of inflation in order to be able to lead the economic agents[1] to make optimal decisions both for themselves and for the society in general.

3.2.1   Determination of monetary policy by a central bank

Many countries have adopted monetary policy regimes in recent years, which specifically target inflation as a key economic variable that should be managed, primarily, using the tool of central bank interest rates (the rates at which banks themselves can borrow). In a situation where inflation is on the rise, it is common for central banks to consider raising interest rates to restrict the availability of money and the speed of its movement through the economy. Clearly, then, central bankers have a need for accurate inflation statistics. In the UK, the Monetary Policy Committee of the Bank of England meets every month to discuss inflation and its future expectations to decide the level of interest rates and, more recently, to determine an appropriate approach to quantitative easing.[2]

3.2.2   Changing of provisions for private pensions

Pensions are structured to provide retirees with an income throughout their retirement; however, it is common for prices to change considerably over this period. If inflation were at the Bank of England's target of 2% per annum and a retiree were to live for 15 years following the end of his/her working life, then the prices would rise by $(1.02)^{15} \approx 34.6\%$ over such a period. To maintain the value of the pension, yearly increases would be needed to match the rise in the general price level. In either case, it is clear that knowledge of changing prices is important if pensioners are to be able to maintain their purchasing power throughout their retirement.

3.2.3   Changes in amounts paid over long-term contracts

Contracts are often designed to run over several years, with structured payments throughout their lives. If both parties agree to a schedule of payments prior to the start of the contract, it is possible that one may benefit and the other may lose, due to unexpected changes in price levels. This is especially true for contracts of long

---

[1] An economic agent is a person or organisation that has an effect on the economy of a country by buying, selling or investing.

[2] Quantitative easing is when the amount of money in the economy is increased by a country's central bank at a time of low interest rates. This is used to stimulate economic growth.

duration, for example, in infrastructure or aerospace projects. In order to account for this issue, it is common to include in such contracts a provision for changing the scheduled payments according to inflation. Without such clauses, it is possible that the profitability of a project could be heavily influenced by the inflation rate, causing potential difficulties in negotiating long-term contracts. It should be noted that depending on the contract, it may be necessary to define an appropriate inflation measurement for the specific area of that contract, for example, an index appropriate to the inputs required in a specific industry.

3.2.4   Changes in rail fares and other goods

The government uses inflation measures to inform a number of price changes as do other economic agents. The idea is that where prices are raised by an amount related to inflation, this is not out of line with the overall inflation experience. In the UK, rail fares[3] can rise by the July RPI increase plus an amount as high as 8%, while on the average, fares must increase by the same number plus 1%. It is, therefore, possible that train companies could use such a requirement to increase their revenue far in advance of the inflation rate; however, being commercial companies, they must weigh this against the response of their customers. Train fares are an example of how inflation is used explicitly in the pricing process.

3.2.5   Evaluating investment decisions

Both private and public sector institutions are likely to consider their expectations of inflation in making long-term financial decisions as changes in the value of money might have a significant effect on the real costs of financing a project. If inflation far outstrips the interest required on a loan, then in real terms, this diminishes the value of the principal to be repaid in real terms. However, the opposite could also be true and in the rare case where the value of money increases over time (deflation), it can become much more onerous to pay back loans.

An example: consider a widget company that borrows £100 at an interest rate of 1%; we assume inflation is running at a rate of 3% and that the prices of widgets increase in line with the overall inflation. If the price of a widget is initially £10, then the company is effectively borrowing 10 widgets worth of money. Over the course of the following year, the debt increases to £101 while the price of widgets increases to £10.30, so the debt is now equivalent to 9.8 widgets.

If inflation outstrips the interest rate, then the amount owed on the debt will become cheaper in real terms (this is the amount the company has to sell in order to cover the debt). For this reason, it is important for those making financial decisions to be aware of inflation.

---

[3] For more information see http://www.rail-reg.gov.uk/server/show/nav.2726

3.2.6   Inputs to economic research and analysis

Statistics about the state of the economy are an important input to economic research and analysis. In order for professional economists to understand and interpret the state of the economy and to test and validate their theories in the real world, it is necessary to acquire a range of information about the status of the economy, including the purchasing power of money. Although this does not mean that economists will be able to necessarily prevent financial crises, such as that experienced in 2008, it does provide them with an opportunity to better understand the conditions that create such crises and their impact on the economy. Reliable measures of inflation are, therefore, of critical importance to economists.

3.2.7   Index-linked debt

Debt issued by the government is often directly linked to a measure of inflation, thus inflation-proofing the payments to the holders of the debt. This is a key consideration of those who hold the debt, and therefore can be seen as an important part of determining how expensive it is for the government to issue debt. In the UK, the government issues its debt linked to the RPI rather than the CPI, as the effect of the gap between the two measures of inflation could significantly affect the demand for government debt and thus impact on the ability of the government to raise money via an issue of long-term debt [3]. As inflation changes, the cost of servicing this debt will also change, and so it can be seen that inflation can have a key role to play in the determination of fiscal as well as monetary policy.

3.2.8   Tax allowances

Tax allowances in the UK are often increased in line with inflation measures, and so it is important that these are increased appropriately in order to ensure that they do not unduly affect either the economic gains of individuals from employment or the government's fiscal situation. A reliable measure of inflation is, therefore, of critical importance in maintaining the appropriateness of the tax system, and a failure in this aspect could have far-reaching implications for the economic behaviour of individuals and governments.

3.2.9   Targets for stability of the economy in an international context

Information about inflation statistics has also been used in an international political context, most notably as a condition for entry into the European Union under the 1992 Maastricht Treaty, which required that a country's inflation measure, as defined by the European Union, fits within a required range. Inflation is also measured by other international organisations such as the World Bank. Inflation measures have a role in an increasingly globalised economy, which goes beyond considerations of how price changes are affecting the consumption experiences of a country's citizens.

3.3   Chapter summary

In this chapter, we have explored what we mean by inflation and explained why measuring it is so important. We defined inflation as a rise in the general price level or a fall in the value of money and discussed the need to define the group of people carrying out transactions and the type of economic transactions we wish to include within the scope of inflation measure.

In Section 3.2, we discussed some of the varied uses of inflation statistics, from high-level politics to industrial contracts and pension provision. Though this list is far from exhaustive, it does highlight the need from a range of sources for a robust and well-understood measure of a vitally important economic statistic.

Inflation is discussed further in Chapter 8, where we explore how the CPI is calculated. Chapter 4 introduces the basic variables that are used to estimate inflation – price and quantity.

References

1. See Parkin, M. (2008) Inflation, in *The New Palgrave Dictionary of Economics* (eds S.N. Durlauf and L.E. Blume), Palgrave Macmillan, 6 September, 2013, doi: 10.1057/9780230226203.0791.

2. See Fisher, I.N. (1956) *My Father Irving Fisher*, Comet Press, New York, p. 235.

3. See UK Debt Management Office (2011) CPI Linked Gilts: A Consultation Document, http://www.dmo.gov.uk/index.aspx?page=Gilts/consultation_papers (accessed 13 January 2015).

Exercise B

These questions are intended to encourage you to think more broadly about the topics discussed in the first three chapters and how they can be applied in practice. Formal solutions are not provided for these questions; however, some brief supporting information for this exercise is available as part of the on-line content associated with this book.

B.1 Identify the benefits and taxes adjusted by inflation measures.

B.2 Given the potential uses of inflation statistics discussed in this chapter, what are the possible implications of having inaccurate or wrong inflation index numbers?

B.3 If you were running a construction business, how might you use index numbers?

B.4 Look for news stories that report inflation statistics; what significance do they place on the numbers they report?

B.5 If inflation is much higher in one area of the UK than the other, how might it influence some of the uses of inflation described in this chapter?

4

Introducing price and quantity

The applications of inflation measures are wide and varied; it is therefore appropriate that significant effort is put into the design and construction of these measures. Since inflation is the rate of change in the general level of prices, it is the latter on which we shall focus; once we have an effective way of measuring the general level of prices, it is a simple matter to calculate how they change. A corresponding measure of the change in quantities bought is also of importance as we shall see later in the book.

In this chapter, we introduce a mathematical representation of prices and quantities and begin to explore how to summarise price and quantity change for collections of goods and services.

4.1 Measuring price change

Before trying to measure price changes for the entire consumer economy, it makes sense to begin modestly with just a single item; by considering changes in price for a single item and then for a small collection of items, we can gain an understanding of how changes in individual prices might influence the general level of prices.

Consider the change in price of a single item between two time periods – 0 and t; let p^t represent the price of a consumer item at time period t and p^0 the price at time period 0. The item could be anything found in a normal shopping basket, for example, a bottle of washing-up liquid. The time period 0 might be January 2014 and the time period t might be July 2014. The change in price can be expressed as the ratio of prices at the two time periods; we call this a price relative:

$$R^{0,t} = \frac{p^t}{p^0} \tag{4.1}$$

This is very similar to the general expression, which was introduced in Chapter 2 for an index number. For a single item, the index number that quantifies the change

A Practical Introduction to Index Numbers, First Edition. Jeff Ralph, Rob O'Neill and Joe Winton.
© 2015 John Wiley & Sons, Ltd. Published 2015 by John Wiley & Sons, Ltd.
Companion Website: http://www.wiley.com/go/ralph/index_numbers

in price is simply the price relative:

$$I^{0,t} = 100 \times R^{0,t} = 100 \times \frac{p^t}{p^0} \qquad (4.2)$$

If, for example the price of a bottle of washing-up liquid is £1.30 in January 2014 and £1.40 in July 2014, then we can use the formula from Chapter 2 for calculating the rate of change in price, or the rate of inflation (as a percentage), as follows:

$$I^{0,t} - 100 = 100 \times (R^{0,t} - 1) = 100 \times \left(\frac{p^t}{p^0} - 1\right) = 100 \times \left(\frac{1.40}{1.30} - 1\right) = 8.33\%$$

We would then say that the rate of inflation between July 2014 and January 2014 is 8.33% based on the change in the price of a bottle of washing-up liquid.

Now, consider a basket of N consumer items and let p_i^t be the price of item i in time period t (the current period) and let p_i^0 be the price of item i in time period 0 (the base period, which we assume occurs at some point in the past). The change in price of this item between the base period, 0, and the current period, t, is then calculated as the **price relative** ($R_i^{0,t}$):

$$R_i^{0,t} = \frac{p_i^t}{p_i^0}$$

A price relative is a specific form of the simple relative index that we saw in Chapter 2; we calculate the change in price from the base period. This allows for an easy interpretation of the percentage change of the price between the base and current periods as we can estimate the percentage change by subtracting 1 from the value of the index and multiplying by 100:

$$\% \text{ change in price} = 100 \times (R_i^{0,t} - 1)$$

This is a simpler calculation than, for example, having to identify the percentage increase of a good, which changes price from £21 in the base period to £24 in the current period.

Example 4.1 Table 4.1 presents the price of five drinks served in a particular bar on the same day each year for 3 years.
    Calculate the price relative for:

a) A pint of bitter in 2014 with 2012 as the base period

b) A vodka (single) in 2013 with 2012 as the base period

For each of the price relatives, describe the change in price between the current and base periods.

Solution For both parts, we need to calculate a price relative

$$R_i^{0,t} = \frac{p_i^t}{p_i^0}$$

Table 4.1   Price of drinks on a particular day, 2012–2014.

		Price (£)		
i	Product	2012	2013	2014
1	Bitter (pint)	1.95	1.80	2.00
2	Lager (pint)	2.40	2.65	2.60
3	Wine (large glass)	3.50	3.50	3.50
4	Vodka (single)	2.20	2.25	2.30
5	Lemonade (half pint)	0.80	0.90	0.95

a) Item 1 (Bitter – pint). The current period is 2014 and the base period is 2012, $p_1^{2014} = 2.00$ and $p_1^{2012} = 1.95$ so:

$$R_1^{2012,2014} = \frac{p_1^{2014}}{p_1^{2012}} = \frac{2.00}{1.95} = 1.026 \quad (3 \text{ d.p.})$$

Between 2012 and 2014, the price of a pint of bitter increased by 2.6%.

b) Item 4 (Vodka – single). The current period is 2013 and the base period is 2012, $p_4^{2013} = 2.25$ and $p_4^{2012} = 2.20$ so:

$$R_4^{2012,2013} = \frac{p_4^{2013}}{p_4^{2012}} = \frac{2.25}{2.20} = 1.023 \quad (3\,\text{d.p.})$$

Between 2012 and 2013, the price of a single vodka increased by 2.3%.

The same considerations apply to the quantities; we can define quantity relatives and calculate percentage changes as for prices. Let q_i^t be the quantity of item i in the current period and let q_i^0 be the quantity of item i in the base period. The change in quantity of this item between the base period, 0, and the current period, t, is called the **quantity relative** ($\tilde{R}_i^{0,t}$):

$$\tilde{R}_i^{0,t} = \frac{q_i^t}{q_i^0} \tag{4.3}$$

The percentage change in the quantity of the ith commodity between the time period 0 and time period t is given by:

$$\% \text{ change in quantity} = 100 \times (\tilde{R}_i^{0,t} - 1)$$

4.2   Simple, un-weighted indices for price change

We have seen how to calculate the change in price of a single item; however, in order to measure price changes for an entire economy, we need some way of combining measures of price change for groups of items.

Deciding on a formula to use when combining changes in price (or changes in quantity) for a collection of commodities is not a simple matter; in fact, it presents a major challenge. There are many ways to combine a set of measurements into one summary figure, which means that there are many ways to combine measures of price change. This section introduces some of the simpler ways. In doing so, we will only consider prices; however, exactly the same considerations apply to quantities.

Much of the index number theory uses the concept of a **'basket of goods'**; this was described briefly in Chapter 1. A basket of goods and services describes a fixed set of commodities for which we measure changes in price and quantity. Imagine a shopping basket; into that basket, we put all of the items that we buy each month. If we can then measure the price of these items each month and calculate the change in the general level of prices for the basket, we have a measure of inflation. This is the approach used in the calculation of the Consumer Prices Index (CPI). In reality, the items or commodities include both consumer products and services. For example, the basket of goods for the UK CPI includes everyday items such as milk, bread, bacon and beer as well as other important purchases such as household gas and electricity bills, big ticket purchases such as fridges, cars and TVs and even holidays, internet dating services and concert tickets.[1] The basket then includes a range of physical and non-physical goods and services of varied sizes and values. Some are consumed immediately, while for others, the services they provide are 'consumed' over significant periods of time.

In practice, the basket of goods and services is a selection of items, chosen to be representative of the huge range of consumer items available. The basket concept is also used beyond the measurement of inflation; composite measures use 'baskets' of dimensions and indicators as we saw in Chapter 1.

4.2.1   Simple price indices

Five different measures of the summary price change for a basket of goods and services are defined below. They only contain the prices of the goods and services and not the quantities; for this reason, they are called 'simple'. Each formula is a price index for a selection of n goods and services between the base period 0 and the current period t; all commodities in the basket are common to both periods 0 and t. Note that the term 'price index' refers to a mathematical formula, which combines prices (and sometimes quantities) to produce price index numbers. In this chapter, we make no comment on the differences between these measures of price change; this will be covered in more detail in Chapter 13.

---

[1] The specification of the CPI basket of goods can be found here: http://www.ons.gov.uk/ons/rel/cpi/cpi-rpi-basket/index.html

Table 4.2   Price of drinks on a particular day, 2012–2013.

i	Product	p_i^{2012}	p_i^{2013}	$R_i^{2012,2013}$
1	Bitter (pint)	1.95	1.80	0.923
2	Lager (pint)	2.40	2.65	1.104
3	Wine (large glass)	3.50	3.50	1.000
4	Vodka (single)	2.20	2.25	1.023
5	Lemonade (half pint)	0.80	0.90	1.125

The Carli[2] price index is an arithmetic mean of price relatives

$$P_{\text{Carli}}^{0,t} = 100 \times \frac{1}{n} \sum_{i=1}^{n} R_i^{0,t} \tag{4.4}$$

The Jevons price index is a geometric mean of price relatives

$$P_{\text{Jevons}}^{0,t} = 100 \times \left(\prod_{i=1}^{n} R_i^{0,t} \right)^{1/n} \tag{4.5}$$

The Dutot price index is a ratio of the arithmetic mean of prices

$$P_{\text{Dutot}}^{0,t} = 100 \times \frac{\frac{1}{n} \sum_{i=1}^{n} p_i^t}{\frac{1}{n} \sum_{i=1}^{n} p_i^0} \tag{4.6}$$

The Harmonic price index is a harmonic mean of price relatives

$$P_{\text{Harmonic}}^{0,t} = 100 \times \frac{n}{\sum_{i=1}^{n} \frac{1}{R_i^{0,t}}} \tag{4.7}$$

The Carruthers, Selwood, Ward and Dalen (CSWD) price index is the geometric mean of the Carli and Harmonic price indices

$$P_{\text{CSWD}}^{0,t} = \sqrt{P_{\text{Carli}}^{0,t} \times P_{\text{Harmonic}}^{0,t}} \tag{4.8}$$

Example 4.2 Table 4.2 presents part of the data from Example 4.1: the price of five drinks served in a particular bar on the same day each year for 2 years alongside the price relative for each item in 2013 with 2012 as the base period.

For these five items (with 2013 as the current period and 2012 as the base period), calculate:

a) the Carli price index number

b) the Jevons price index number

---

[2] It is common practice in index numbers for index formulae to be named after the people who first introduced them.

c) the Dütot price index number

d) the Harmonic price index number

e) the CSWD price index number

Solution

a) The Carli index is an arithmetic mean of price relatives

$$P^{0,t}_{Carli} = 100 \times \frac{1}{n}\sum_{i=1}^{n} R^{0,t}_i$$

$$P^{2012,2013}_{Carli} = 100 \times \frac{1}{5}\sum_{i=1}^{5} R^{2012,2013}_i$$

$$= 100 \times \frac{0.923 + 1.104 + 1.000 + 1.023 + 1.125}{5} = 103.5 \ (1 \ \text{d.p.})$$

b) The Jevons Index is a geometric mean of price relatives

$$P^{0,t}_{Jevons} = 100 \times \left(\prod_{i=1}^{n} R^{0,t}_i\right)^{1/n}$$

$$P^{2012,2013}_{Jevons} = 100 \times \left(\prod_{i=1}^{5} R^{2012,2013}_i\right)^{1/5}$$

$$= 100 \times (0.923 \times 1.104 \times 1.000 \times 1.023 \times 1.125)^{1/5}$$

$$= 103.2 (1 \ \text{d.p.})$$

c) The Dütot Price Index is a ratio of the arithmetic mean of prices

$$P^{0,t}_{Dutot} = 100 \times \frac{\frac{1}{n}\sum_{i=1}^{n} P^t_i}{\frac{1}{n}\sum_{i=1}^{n} P^0_i}$$

$$P^{2012,2013}_{Dutot} = 100 \times \frac{\frac{1}{5}\sum_{i=1}^{n} P^{2013}_i}{\frac{1}{5}\sum_{i=1}^{n} P^{2012}_i} = 100 \times \frac{\sum_{i=1}^{n} P^{2013}_i}{\sum_{i=1}^{n} P^{2012}_i}$$

$$= 100 \times \frac{(1.80 + 2.65 + 3.50 + 2.25 + 0.90)}{(1.95 + 2.40 + 3.50 + 2.20 + 0.80)}$$

$$= 100 \times \frac{11.10}{10.85} = 102.3 (1 \ \text{d.p.})$$

d) The Harmonic Price Index is a harmonic mean of price relatives

$$P^{0,t}_{Harmonic} = 100 \times \frac{n}{\sum_{i=1}^{n} \frac{1}{R^{0,t}_i}}$$

$$P_{\text{Harmonic}}^{2012,2013} = \frac{500}{\sum_{i=1}^{n} \frac{1}{R_i^{2012,2013}}} = \frac{500}{\left(\frac{1}{0.923} + \frac{1}{1.104} + \frac{1}{1.000} + \frac{1}{1.023} + \frac{1}{1.125} \right)}$$

$$= \frac{500}{4.9} = 103.0 \ (1 \ \text{d.p.})$$

e) The CSWD Price Index is a geometric mean of the Carli and Harmonic price indices

$$P_{\text{CSWD}}^{0,t} = \sqrt{P_{\text{Carli}}^{0,t} \times P_{\text{Harmonic}}^{0,t}}$$

$$P_{\text{CSWD}}^{2012,2013} = \sqrt{P_{\text{Carli}}^{2012,2013} \times P_{\text{Harmonic}}^{2012,2013}} = \sqrt{103.5 \times 1.30} = 103.2 \ (1 \ \text{d.p.})$$

Note that in the examples above, the estimates of inflation range from 2.3 to 3.5% with the same price data. These differences may not cause immediate concern; however, if this same range of values was suggested for the headline inflation results, the variation in possible values of inflation would have significant consequences given the range of uses of inflation as discussed in Chapter 3. This simple example shows that the choice of index formula is an important consideration when deciding how to construct a measure of inflation.

4.2.2   Simple quantity indices

The same five indices can be used to combine measures of quantity change by replacing price data with quantity data.

The Carli quantity index is an arithmetic mean of quantity relatives

$$Q_{\text{Carli}}^{0,t} = 100 \times \frac{1}{n} \sum_{i=1}^{n} \tilde{R}_i^{0,t} \tag{4.9}$$

The Jevons quantity index is a geometric mean of quantity relatives

$$Q_{\text{Jevons}}^{0,t} = 100 \times \left(\prod_{i=1}^{n} \tilde{R}_i^{0,t} \right)^{1/n} \tag{4.10}$$

The Dutot quantity index is a ratio of the arithmetic mean of quantities

$$Q_{\text{Dutot}}^{0,t} = 100 \times \frac{\frac{1}{n} \sum_{i=1}^{n} q_i^t}{\frac{1}{n} \sum_{i=1}^{n} q_i^0} \tag{4.11}$$

The Harmonic quantity index is a harmonic mean of quantity relatives

$$Q_{\text{Harmonic}}^{0,t} = 100 \times \frac{n}{\sum_{i=1}^{n} \frac{1}{\tilde{R}_i^{0,t}}} \tag{4.12}$$

The CSWD quantity index is a geometric mean of the Carli and Harmonic quantity indices

$$Q^{0,t}_{\text{CSWD}} = \sqrt{Q^{0,t}_{\text{Carli}} \times Q^{0,t}_{\text{Harmonic}}} \tag{4.13}$$

If we calculate the quantity index numbers for these five quantity indices using quantity data, we would also see a range of values. For the corresponding quantity data, the different index numbers would represent the change in the 'general level of quantities' between the two time periods. The relationship between price and quantity is explored in the next section and in the following chapter.

4.3   Price, quantity and value

The relationship between price, quantity and value is central to index number theory. These three variables are linked in the following way. For any item the value can be calculated as the unit price of that item multiplied by the quantity:

$$\text{value} = \text{price} \times \text{quantity} \tag{4.14}$$

For an item i at *time period t*, the value is calculated as the price of one unit of the item at time period 't' multiplied by the quantity of the item sold at time 't':

$$v^t_i = p^t_i q^t_i \tag{4.15}$$

For a collection of n items (a basket of n goods and services), the total value at period t is the sum of the value of the items in the basket:

$$V^t = \sum_{i=1}^{n} p^t_i q^t_i \tag{4.16}$$

Depending on the setting, value can also be thought of as expenditure, turnover or cost; these terms are interchangeable in discussions of value and will be used throughout this book.

The change in value of a set of items between two time periods (say 0 and t) is then expressed as the ratio of values:

$$V^{0,t} = \frac{v^t}{v^0} = \frac{\sum_{i=1}^{n} p^t_i q^t_i}{\sum_{i=1}^{n} p^0_i q^0_i} \tag{4.17}$$

This is useful as a quick calculation of change; again this can be calculated as a percentage by multiplying ($V^{0,t} - 1$) by 100.

Where the total value changes between two time periods, the formula shows that the change could be driven by a change in prices, or by a change in quantities or a combination of both. For example, if the price of your weekly shop goes up, is it because you bought more of some items, because some of the items you bought cost more or is it some combination of the two? In fact, it is often the case that changes in price are negatively correlated with changes in quantity, that is when the price of an item goes up, less of it is bought.

4.4   Example – Retail Sales Index

The Retail Sales Index (RSI) produced by the Office for National Statistics in the UK measures the growth in value of sales for different retail sectors. Table 4.3 shows the growth in the value of retail sales from June 2012 to June 2013 for three retail sectors: 'Predominantly Food Stores', 'Automotive Fuel' and 'Textiles, Clothing and Footwear'.

Table 4.3   Value of retail sales[a] June 2013.

Percentage change on the same month a year ago	
	Value
Predominantly food stores	3.0
Automotive fuel	2.6
Textiles, clothing, and footwear	4.2

[a] Non-seasonally adjusted.

From Table 4.3, it is clear that there has been an increase in the value of retail sales (the amount of money spent) for all three sectors, but the value alone cannot explain the underlying reasons for the growth. Is this growth due to a change in prices or is it due to a change in the quantities sold? What is more likely is that it is some combination of the two. Price and quantity need not move in the same direction. As price decreases, it is expected that demand will grow and the quantity of items bought will increase; the way in which the two variables change is important in economic theory.

Alongside the *value* of retail sales, the ONS RSI release also includes measures of the growth in the *volume* of sales for different retail sectors. Table 4.4 shows this information alongside the implied growth in price and the growth in value for the three sectors in Table 4.3. The volume of sales is a measure of the overall quantity of retail sales. Note that 'volume' means the same as 'quantity';a quantity index is sometimes referred to as a volume index.

Table 4.4   Retail sales[a] 2013.

Percentage change on the same month a year ago			
	Value	Volume	Price
Predominantly food stores	3.0	−0.4	3.4
Automotive fuel	2.6	−0.4	3.0
Textiles, clothing and footwear	4.2	2.9	1.3

[a] Non-seasonally adjusted.
Source: Office for National Statistics.

When we combine the growth in value with the growth in both price and volume, we see a much clearer picture. For all three sectors, the amount of money being spent has risen in the 12-month period; however, in the case of both 'Food' and 'Fuel', the quantity bought has actually decreased and the growth in value is being driven by a larger increase in prices.

This example illustrates one of the key discussions in index number theory: The Index Number Problem.

The Index Number Problem[3]

'How to combine the relative changes in the prices and quantities of various products into

(i) a single measure of the relative change of the overall **price level** and

(ii) a single measure of the relative change of the overall **quantity level**'.

Expressed mathematically, this is asking whether we can define price and quantity indices, which when combined together by some mathematical function provide us with the change in value:

$$V^t = f(P^{0,t}, Q^{0,t}) \qquad (4.18)$$

4.5 Chapter summary

Many of the developments and issues in the theory of index numbers have focused on how to combine prices or price relatives into an overall change in price for a collection of different items. The relationship between price and quantity is central to this problem and will be discussed further throughout this book.

This chapter has introduced us to the topic of measuring price and quantity change across multiple goods, and we have seen that by combining individual prices (quantities) in different ways, we get a different measure of the price (quantity) movement for a selection of goods. In later chapters, we will discuss the relative merits of each index formula and introduce some approaches to choosing between them.

We have also introduced the Index Number Problem and suggested that measures of price and quantity change should combine in some way to reveal something about movements in the overall value of an economic series.

---

[3] OECD Glossary of Statistical terms: http://stats.oecd.org/glossary/detail.asp?ID=5639

Exercise C

Solutions to this exercise can be found in Appendix D.

C.1 Table C.1 presents the price of five cakes over a 3-month period:

Table C.1   Cake prices March-May.

	Price (£)		
	March p^0	April p^1	May p^2
Chocolate chip cookie	1.20	1.50	1.35
Lemon drizzle cake	1.30	1.30	1.25
Caramel shortbread	1.00	1.20	1.00
Flapjack	1.10	1.20	1.25
Rocky road	1.90	1.95	2.10

For this selection of cakes in April and May, calculate (with March as the base period):

a. the Carli price index numbers

b. the Jevons price index numbers

c. the Dutot price index

d. the Harmonic price index

e. the CSWD price index

C.2 Table C.2 presents the number of hours of tuition for two subjects provided by a music company in 2 years, 2005 and 2010, along with the price for 1-h tuition.

a. Calculate a *price* relative for each item in 2010 with 2005 as the base period.

b. Calculate a *quantity* relative for each item in 2010 with 2005 as the base period.

c. Comment on the price and quantity relatives of the two items and any relationships between the items.

Table C.2   Price and quantity of music lessons.

	Price (£)		Quantity	
	2005 (p^0)	2010 (p^t)	2005 (q^0)	2010 (q^t)
Piano lesson	15	22	268	186
Guitar lesson	15	16	208	269

C.3 Table C.3 presents the prices for five varieties of wine in two periods, 2013 and 2014.

a. Calculate the Dutot price index number for 2014 with 2013 as the base period

b. Calculate the Jevons price index number for 2014 with 2013 as the base period

c. Comment on the difference between your answers to a. and b.

Table C.3   Price of wine 2013–2014.

	Price (£)	
	2013 (p^0)	2014 (p^t)
Vintage champagne	180.00	250.00
European red	6.70	7.30
New World red	5.00	5.50
European white	5.20	5.70
New World white	5.90	6.50

5

Laspeyres and Paasche indices

In Chapter 4, simple price relatives were calculated and combined using a selection of simple index formulae. In this chapter, the problem of how to combine price changes is explored further. We will also further examine the relationship between price, quantity and value and introduce two important index number formulae: those of Laspeyres and Paasche.

Table 5.1 shows the prices of three fruits in 2012 and 2013 along with their calculated price relatives, using the formula introduced in Chapter 4.[1] The price of a pineapple rose by 60% between 2012 and 2013 compared to a rise of 10% for bananas and a relatively small change of 5% in the price of apples. We can use some of the simple index formulae that we saw in Chapter 4 to combine this information; the results are displayed in Table 5.2.

Depending on the formula that we use to combine the price changes, the price of the basket of fruit has risen by 23, 25 or 38%. Each of these ways of combining price change is a reasonable measurement of the overall price change.

The price change of apples and bananas was small compared to the change in the price of a pineapple, as a result the simple price indices (Carli, Dütot and Jevons) are heavily influenced by the large price change of the pineapples; but is this what we want? Are the large price movements of pineapples just as important to this basket as the small price movements in apples? Let us now introduce some additional information; Table 5.3 shows the quantity of each fruit sold in 2012. In the base period, 220 apples and 230 bananas were sold compared to only 50 pineapples. Taking into account this new information, are the price movements of pineapples just as important as the price movements in apples and bananas when measuring inflation for this basket?

---

[1] The prices, quantities and the sources in this chapter have been invented just to illustrate the effects of price and quantity changes.

A Practical Introduction to Index Numbers, First Edition. Jeff Ralph, Rob O'Neill and Joe Winton.
© 2015 John Wiley & Sons, Ltd. Published 2015 by John Wiley & Sons, Ltd.
Companion Website: http://www.wiley.com/go/ralph/index_numbers

Table 5.1   Fruit prices.

	2012 price (p^0)	2013 price (p^1)	Price relative ($R^{0,1}$)
Apples	£0.40	£0.42	1.050
Bananas	£0.50	£0.55	1.100
Pineapple	£1.25	£2.00	1.600

Table 5.2   Selected fruit price indices.

(Base = 2012)	
	Price index for 2013 ($P^{0,2013}$)
Carli	125.0
Dütot	138.1
Jevons	122.7

Table 5.3   Fruit prices and quantities.

	2012 price (p^0)	2013 price (p^1)	2012 quantity (q^0)
Apples	£0.40	£0.42	220
Bananas	£0.50	£0.55	230
Pineapple	£1.25	£2.00	50

It is reasonable to conclude that not all items in our basket are equally important. When this is the case, we would like to apply a weighting to take this into account. In the next section, we use additional information that is available about the items in our basket to determine these weights.

5.1   The Laspeyres price index

The Laspeyres price index uses quantities from the base period to help determine the importance of each item in the basket. The **Laspeyres price index** is defined as:

$$P^{0,t}_{\text{Laspeyres}} = 100 \times \frac{\sum_{i=1}^{n} p_i^t q_i^0}{\sum_{i=1}^{n} p_i^0 q_i^0} \tag{5.1}$$

The Laspeyres price index measures the change in value between two periods, with the quantities fixed in the base period. By fixing the quantities, the Laspeyres price index measures the effect of the change in prices on the change in value of the basket.

It answers the question: how has the cost of buying this selection of goods changed if I buy the same quantity of each good in the current period as I did in the base period?

Using the information from Table 5.3, we can calculate a Laspeyres price index number for the basket of fruit in 2013 with 2012 as the base period:

$$P_{\text{Laspeyres}}^{2012,2013} = 100 \times \frac{\sum_{i=1}^{3} p_i^{2013} \times q_i^{2012}}{\sum_{i=1}^{3} p_i^{2012} \times q_i^{2012}}$$

$$= 100 \times \frac{p_{\text{Apples}}^{2013} \times q_{\text{Apples}}^{2012} + p_{\text{Bananas}}^{2013} \times q_{\text{Bananas}}^{2012} + p_{\text{Pineapples}}^{2013} \times q_{\text{Pineapples}}^{2012}}{p_{\text{Apples}}^{2012} \times q_{\text{Apples}}^{2012} + p_{\text{Bananas}}^{2012} \times q_{\text{Bananas}}^{2012} + p_{\text{Pineapples}}^{2012} \times q_{\text{Pineapples}}^{2012}}$$

$$= 100 \times \frac{(0.42 \times 220) + (0.55 \times 230) + (2.00 \times 50)}{(0.40 \times 220) + (0.50 \times 230) + (1.25 \times 50)}$$

$$= 100 \times \frac{319}{266} = 120.1$$

Using the Laspeyres price index leads to a price change of around 20% for the basket of fruit. By incorporating information about the quantity of each item in the base period, the Laspeyres price index gives a measure of price change, which is less influenced by the large price change in pineapples than the un-weighted, simple alternatives such as the Carli, Dütot and Jevons indices (Figure 5.1).

5.2 The Paasche price index

By fixing the quantities in the base period, the Laspeyres price index measures the effect of price changes on the change in value of the basket if the base period quantities are used throughout. However, it is equally valid to hold the quantities constant at their *current* period level. This answers the question: how has the cost of buying this selection of goods changed if I had bought the same quantity of each good in the base period as I do in the current period? This may initially seem less intuitive than using the base period, but there is no obvious theoretical reason for us to choose the quantities from the base period rather than from the current period.

The **Paasche price index** does just this; it uses quantities from the current period to help determine the importance of each item in the basket. The Paasche price index is defined as:

$$P_{\text{Paasche}}^{0,t} = 100 \times \frac{\sum_{i=1}^{n} p_i^t q_i^t}{\sum_{i=1}^{n} p_i^0 q_i^t} \tag{5.2}$$

Table 5.4 shows the complete price and quantity information for the basket of fruit in 2012 and 2013. In 2013, the number of apples sold has risen to 300 but there have been falls in the number of bananas and pineapples sold. In 2012, a similar number of apples and bananas were sold; however, in 2013, apples were by far the highest selling fruit. A Paasche price index number for this basket of fruit in 2013 with 2012

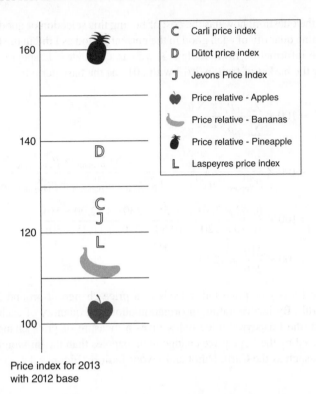

Figure 5.1   Measures of price change for fruit between 2012 and 2013.

as the base period can be calculated as follows:

$$P_{\text{Paasche}}^{2012,2013} = 100 \times \frac{\sum_{i=1}^{3} p_i^{2013} \cdot q_i^{2013}}{\sum_{i=1}^{3} p_i^{2012} \cdot q_i^{2013}}$$

$$= 100 \times \frac{p_{\text{Apples}}^{2013} \cdot q_{\text{Apples}}^{2013} + p_{\text{Bananas}}^{2013} \cdot q_{\text{Bananas}}^{2013} + p_{\text{Pineapples}}^{2013} \cdot q_{\text{Pineapples}}^{2013}}{p_{\text{Apples}}^{2012} \cdot q_{\text{Apples}}^{2013} + p_{\text{Bananas}}^{2012} \cdot q_{\text{Bananas}}^{2013} + p_{\text{Pineapples}}^{2012} \cdot q_{\text{Pineapples}}^{2013}}$$

$$= 100 \times \frac{(0.42 \times 300) + (0.55 \times 170) + (2.00 \times 30)}{(0.40 \times 300) + (0.50 \times 170) + (1.25 \times 30)}$$

$$= 100 \times \frac{279.5}{242.5} = 115.3$$

Using a Paasche price index to measure the overall change in price for this basket of fruit results in a smaller price change than any of the other formulae we have used. This is because the use of current quantities gives more importance to the small

Table 5.4   Fruit prices and quantities.

	2012 price (p^0)	2013 price (p^1)	2012 quantity (q^0)	2013 quantity (q^1)
Apples	£0.40	£0.42	220	300
Bananas	£0.50	£0.55	230	170
Pineapple	£1.25	£2.00	50	30

change in the price of apples and is less influenced by the large change in price of the pineapples than the un-weighted indices and even the Laspeyres price index. Note that this makes some sense; as pineapples have increased in price, the demand has dropped off noticeably, which leads to us giving less weight to the large price change for pineapples on the other hand demand has increased for apples and so we give more weight to the fruit with the smallest increase in price.

5.3   Laspeyres and Paasche quantity indices

For all of the measures of price change that we are exploring, it is also possible to calculate an equivalent quantity index by swapping the prices and quantities in the formula.

The **Laspeyres quantity index** is calculated as the change in value between the base and current periods, with *prices* fixed in the base period:

$$Q^{0,t}_{\text{Laspeyres}} = 100 \times \frac{\sum_{i=1}^{n} p_i^0 q_i^t}{\sum_{i=1}^{n} p_i^0 q_i^0} \tag{5.3}$$

By fixing the prices, the Laspeyres quantity index measures the effect of quantity changes on the change in value of the basket if the prices remain constant at their base period levels. It answers the question: how does the cost of buying the different selection of goods change if each good has the same price in the current period as in the base period?

Similarly, the **Paasche quantity index** is calculated as the change in value between the base and current periods, with prices fixed at their current period levels:

$$Q^{0,t}_{\text{Paasche}} = 100 \times \frac{\sum_{i=1}^{n} p_i^t q_i^t}{\sum_{i=1}^{n} p_i^t q_i^0} \tag{5.4}$$

By fixing the prices, the Paasche quantity index measures the effect of quantity changes on the change in the value of the basket if the prices remain constant at their current period levels. It answers the question: What is the difference in the cost of buying the two selections of goods if the price in the base period was the same as that in the current period?

Example 5.1   Table 5.5 shows the price of a selection of tools sold by a DIY shop in 2013 and 2014 along with the amount of each tool that was sold in the same periods.

Table 5.5   Price and quantity of tools, 2013 and 2014.

i	Tool	2013 price (p^1)	2014 price (p^0)	2013 quantity (q^1)	2014 quantity (q^0)
1	Screw driver set	£13.00	£15.30	144	141
2	Electric drill	£23.50	£26.70	155	152
3	Step ladder	£21.85	£22.60	150	140
4	Hammer	£18.70	£20.00	151	149
5	Wood saw	£11.50	£13.00	149	141

For the selection of tools in 2014, with 2013 as the base period, calculate:

a) the Laspeyres price index number

b) the Paasche price index number

c) the Laspeyres quantity index number

d) the Paasche quantity index number

Solutions

a)

$$P_{\text{Laspeyres}}^{2013,2014} = 100 \times \frac{\sum_{i=1}^{5} p_i^{2014} \cdot q_i^{2013}}{\sum_{i=1}^{5} p_i^{2013} \cdot q_i^{2013}}$$

$$= 100 \times \frac{\begin{array}{c}(15.3 \times 144) + (26.7 \times 155) + (22.6 \times 150) \\ + (20 \times 151) + (13 \times 149)\end{array}}{\begin{array}{c}(13 \times 144) + (23.5 \times 155) + (21.85 \times 150) \\ + (18.7 \times 151) + (11.5 \times 149)\end{array}}$$

$$= 100 \times \left(\frac{14688.7}{13329.2} \right) = 110.2$$

b)

$$P_{\text{Paasche}}^{2013,2014} = 100 \times \frac{\sum_{i=1}^{5} p_i^{2014} \times q_i^{2014}}{\sum_{i=1}^{5} p_i^{2013} \times q_i^{2014}}$$

$$= 100 \times \frac{\begin{array}{c}(15.3 \times 141) + (26.7 \times 152) + (22.6 \times 140) \\ + (20 \times 149) + (13 \times 141)\end{array}}{\begin{array}{c}(13 \times 141) + (23.5 \times 152) + (21.85 \times 140) \\ + (18.7 \times 149) + (11.5 \times 141)\end{array}}$$

$$= 100 \times \left(\frac{14192.7}{12871.8} \right) = 110.3$$

c)

$$Q_{\text{Laspeyres}}^{2013,2014} = 100 \times \frac{\sum_{i=1}^{5} p_i^{2013} \times q_i^{2014}}{\sum_{i=1}^{5} p_i^{2013} \times q_i^{2013}}$$

$$= 100 \times \frac{\begin{array}{c}(13 \times 141) + (23.5 \times 152) + (21.85 \times 140)\\ + (18.7 \times 149) + (11.5 \times 141)\end{array}}{\begin{array}{c}(13 \times 144) + (23.5 \times 155) + (21.85 \times 150)\\ + (18.7 \times 151) + (11.5 \times 149)\end{array}}$$

$$= 100 \times \left(\frac{12871.8}{13329.2}\right) = 96.6$$

d)

$$Q_{\text{Paasche}}^{2002,2005} = 100 \times \frac{\sum_{i=1}^{5} p_i^{2014} \times q_i^{2014}}{\sum_{i=1}^{5} p_i^{2014} \times q_i^{2013}}$$

$$= 100 \times \frac{\begin{array}{c}(15.3 \times 141) + (26.7 \times 152) + (22.6 \times 140)\\ + (20 \times 149) + (13 \times 141)\end{array}}{\begin{array}{c}(15.3 \times 144) + (26.7 \times 155) + (22.6 \times 150)\\ + (20 \times 151) + (13 \times 149)\end{array}}$$

$$= 100 \times \left(\frac{14192.7}{14688.7}\right) = 96.6$$

5.4   Laspeyres and Paasche: mind your Ps and Qs

In this chapter, Laspeyres and Paasche price and quantity indices have been expressed in terms of prices and quantities. These formulae can be rearranged so that each index is expressed only in terms of data relating to price relatives and value shares. In practice, it is often more useful to express these indices in this form and below we introduce these versions of the indices; proofs of each of these expressions can be found in Appendix A.

5.4.1   Laspeyres price index as a weighted sum of price relatives

The Laspeyres price index can be expressed as a weighted sum of price relatives, where the weight for each item is its value share in the base period. That is:

$$P_{\text{Laspeyres}}^{0,t} = 100 \times \sum_{i=1}^{n} R_i^{0,t} \cdot s_i^0 \tag{5.5}$$

where s_i^0 is the value share of item i in period 0:

$$s_i^0 = \frac{v_i^0}{\sum_{i=1}^{n} v_i^0} \tag{5.6}$$

$R_i^{0,t}$ is the price relative for item i between period *0 and period t:*

$$R_t^{0,t} = \frac{p_i^t}{p_i^0} \tag{5.7}$$

and v_i^0 is the value of item i in time period 0:

$$v_i^0 = p_i^0 \cdot q_i^0 \tag{5.8}$$

This representation of the Laspeyres price index does have an intuitive appeal. The price change for each commodity is multiplied by a weight, which is the proportion of expenditure on that commodity.

5.4.2   Laspeyres quantity index as a weighted sum of quantity relatives

The Laspeyres quantity index can be expressed as a weighted sum of quantity relatives, where the weight for each item is its value share in the base period. That is:

$$Q_{\text{Laspeyres}}^{0,t} = 100 \times \sum_{i=1}^{n} \tilde{R}_i^{0,t} \cdot s_i^0 \tag{5.9}$$

5.4.3   Paasche price index as a weighted harmonic mean of price relatives

The Paasche price index can be expressed as a weighted harmonic mean of price relatives, where the weight for each item is its value share in the current period. That is:

$$P_{\text{Paasche}}^{0,t} = 100 \times \frac{1}{\sum_{i=1}^{n} \frac{s_i^t}{R_i^{0,t}}} \tag{5.10}$$

5.4.4   Paasche quantity index as a weighted harmonic mean of quantity relatives

The Paasche quantity index can be expressed as a weighted harmonic mean of quantity relatives, where the weight for each item is its value share in the current period. That is:

$$Q_{\text{Paasche}}^{0,t} = 100 \times \frac{1}{\sum_{i=1}^{n} \frac{s_i^t}{\tilde{R}_i^{0,t}}} \tag{5.11}$$

Example 5.2   Table 5.6 shows the price and quantity relatives between 2013 and 2014 as well as value shares for each period for a selection of tools. These values have been calculated from the price and quantity data described in Table 5.5 above.[2]

---
[2] We encourage the reader to calculate these index numbers for themselves.

Table 5.6   Price and quantity relatives of tools, 2013–2014 with value shares.

i	Tool	Price relative $R^{2013,2014}$	Quantity relative $R^{2013,2014}$	2013 value share s_i^{2013}	2014 value share s_i^{2014}
1	Screw driver set	1.177	0.979	0.2	0.1
2	Electric drill	1.136	0.981	0.3	0.3
3	Step ladder	1.034	0.933	0.2	0.2
4	Hammer	1.070	0.987	0.2	0.2

For the selection of tools in 2014, with 2013 as the base period and using only the information in Table 5.6, calculate:

a) the Laspeyres price index number

b) the Paasche price index number

c) the Laspeyres quantity index number

d) the Paasche quantity index number

Solutions

a)

$$P_{\text{Laspeyres}}^{2013,2014} = 100 \times \left(\sum_{i=1}^{5} R_i^{2013,2014} \times s_i^{2014} \right)$$

$$= 100 \times ((1.177 \times 0.1) + (1.136 \times 0.3) + (1.034 \times 0.2)$$
$$+ (1.070 \times 0.2) + (1.130 \times 0.1))$$

$$= 110.2$$

b)

$$P_{\text{Paasche}}^{2013,2014} = 100 \times \frac{1}{\sum_{i=1}^{5} \frac{s_i^{2014}}{R_i^{2013,2014}}}$$

$$= 100 \times \frac{1}{\left(\frac{0.2}{1.177} \right) + \left(\frac{0.3}{1.136} \right) + \left(\frac{0.2}{1.034} \right) + \left(\frac{0.2}{1.070} \right) + \left(\frac{0.1}{1.130} \right)}$$

$$= 110.3$$

c)

$$Q_{\text{Laspeyres}}^{2013,2014} = 100 \times \left(\sum_{i=1}^{5} \tilde{R}_i^{2013,2014} \times s_i^{2013} \right)$$

$$= 100 \times ((0.979 \times 0.1) + (0.981 \times 0.3) + (0.933 \times 0.2)$$

$$+ (0.987 \times 0.2) + (0.946 \times 0.1))$$

$$= 96.6$$

d)

$$Q_{\text{Paasche}}^{2013,2014} = 100 \times \frac{1}{\sum_{i=1}^{5} \frac{s_i^{2014}}{R_i^{2013,2014}}}$$

$$= 100 \times \frac{1}{\left(\frac{0.2}{0.979}\right) + \left(\frac{0.3}{0.981}\right) + \left(\frac{0.2}{0.933}\right) + \left(\frac{0.2}{0.987}\right) + \left(\frac{0.1}{0.946}\right)}$$

$$= 96.6$$

5.5   Laspeyres, Paasche and the Index Number Problem

In Chapter 4, we introduced the Index Number Problem:

The Index Number Problem

   'How to combine the relative changes in the prices and quantities of various products into

   (i) a single measure of the relative change of the overall **price level** and

   (ii) a single measure of the relative change of the overall **quantity level**'

                                                           OECD Definition

---

We can represent this in mathematical form:

$$V^{0,t} = f(\, P^{0,t}, Q^{0,t}\,) \tag{5.12}$$

This says that the overall change in value can be decomposed into an overall change in price and an overall change in quantity, combined by a function f.

Laspeyres and Paasche indices provide a neat solution to the Index Number Problem, and it can be shown that:

$$V^{0,t} = P_{\text{Laspeyres}}^{0,t} \times Q_{\text{Paasche}}^{0,t} \tag{5.13}$$

and:

$$V^{0,t} = P_{\text{Paasche}}^{0,t} \times Q_{\text{Laspeyres}}^{0,t} \tag{5.14}$$

A proof of these results can be found in Appendix A.

This relationship becomes very useful as measuring the overall quantity change is difficult, while value and price change are relatively easily measured; we will see this in Chapter 10.

5.6   Laspeyres or Paasche?

In this chapter, we have introduced the price and quantity indices of Laspeyres and Paasche. Both of these price indices are sensible: for both price indices, we fix quantities at some time point and measure the effect of the change in price using the quantities to determine the importance of each item in our basket. This leads us to ask: 'which price index should we use, Laspeyres or Paasche?' The same question applies to the two quantity indices.

The Laspeyres price index has the edge in practical use. Collecting information on quantities is often difficult, time consuming and expensive – or at least more so than prices or values. Since the Laspeyres price index only requires quantity information from one period – the base period – which is then held constant, only new price information is required in each period. If we calculate a Paasche price index, then we require quantity information for every period that we wish to compare with the base period. This makes the Laspeyres price index easier and cheaper to calculate than the equivalent Paasche price index, which will require new price and quantity information in all periods.

The Paasche price index better reflects current buying habits. By using quantity information from the current period to determine the importance of each item, the Paasche price index adapts to changing buying habits by giving the biggest weight to items with a higher expenditure in the current period. Conversely, the Laspeyres price index is at risk of becoming outdated. It is perhaps not surprising that there is a relationship between the price of an item and the amount of that item purchased. In general, as the price of an item increases, the quantity of that item that is sold will decrease. This means that the Laspeyres price index is at risk of giving a large weight to items where there is a big price increase because it does not account for any accompanying decrease in quantity. The Paasche price index is at risk in the opposite context, that is, it may give a small weight to large decreases in price without accounting for a possible accompanying increase in quantity.

An example: Suppose that a supermarket sells two brands of instant coffee. The two brands are very similar, with 'identical' taste, size, country of origin and packaging.

In January, both the price of a jar of Brand A and the price of a jar of Brand B coffee is £2.50, and a supermarket sold 500 jars of each brand.

In February, the price of a jar of Brand A coffee rises to £4.00, but that of Brand B coffee remains at £2.50. As a result, the supermarket sells 100 jars of Brand A and 900 jars of Brand B in February.

We can calculate the Laspeyres and Paasche price index numbers for February with January as the base month:

$$P_{\text{Laspeyres}}^{\text{Jan,Feb}} = 100 \times \frac{\sum p_i^{\text{Feb}} q_i^{\text{Jan}}}{\sum p_i^{\text{Jan}} q_i^{\text{Jan}}} = 100 \times \frac{(4.00 \times 500) + (2.50 \times 500)}{(2.50 \times 500) + (2.50 \times 500)} = 130.00$$

$$P_{\text{Paasche}}^{\text{Jan,Feb}} = 100 \times \frac{\sum p_i^{\text{Feb}} q_i^{\text{Feb}}}{\sum p_i^{\text{Jan}} q_i^{\text{Feb}}} = 100 \times \frac{(4.00 \times 100) + (2.50 \times 900)}{(2.50 \times 100) + (2.50 \times 900)} = 106.00$$

In this case, there is a large difference between the Laspeyres and Paasche index numbers.

Here, the Laspeyres price index gives equal weight to both brands and ignores the fact that in February, Brand B is by far the more popular choice – this could overestimate the overall price change.

On the other hand, the Paasche price index does account for the shift in spending patterns and gives a much higher weight to Brand B. This could underestimate the overall price change as only a small weight is given to the large price increase in Brand A.

As a result, the Laspeyres price index tends to overstate price increases, whereas the Paasche price index tends to understate them when we consider consumer behaviour in response to relative price changes; this is an issue, which we will return to in Chapter 13.

It is also worth noting that since only the prices change in each time period, the series of Laspeyres price index numbers are directly comparable whereas for Paasche price index versions, this is not true as the quantities used also change in each period.

In deciding which index formula to use, most National Statistics Institutes, which have the responsibility for producing measures of price and quantity change, chose the approach that is often the only one that can be calculated in a practical, timely manner. A measure of inflation that reflects the effect of consumers adjusting what they buy in response to the gradual changes in prices will fall between the Laspeyres and Paasche price indices, and in later chapters, we shall explore some other measures of inflation that try to account for consumer behaviour.

In Chapter 12, we shall explore an average of the Laspeyres and Paasche formulae known as the Fisher index as an alternative measure. The Fisher price index is defined as the geometric mean of the Laspeyres and Paasche price indices calculated from the same price and quantity data over the same time periods.

The quote below suggests another alternative; why not provide both Laspeyres and Paasche indices as upper and lower limits of what inflation could be. Obviously, these approaches have the same downside, as the Paasche price index as current period quantity information is unlikely to be available on time to publish these measures quickly.

In principle, instead of averaging the Paasche and Laspeyres indices, the statistical agency could think of providing both (the Paasche index on a delayed basis). This suggestion would lead to a matrix of price comparisons between every pair of periods instead of a time series of comparisons. Walsh [1] noted this possibility: 'In fact, if we use such direct comparisons at all, we ought to use all possible ones'.

Consumer Price Index Manual: Theory and Practice [2]

5.7   A more practical alternative to a Laspeyres price index?

In this chapter, we have suggested that the Laspeyres price index is preferred in practice to its Paasche equivalent because of the difficulty in obtaining timely quantity (or expenditure weight) data. However, even obtaining these data for the base period in a timely manner for regular publication is difficult.

We will now introduce two other index formulae, which are very similar to those of Laspeyres and Paasche and can be used as alternatives. These index formulae will be discussed again in Chapter 12.

The **Lowe price index** is similar to both its Laspeyres and Paasche equivalents, but rather than fixing quantities in the base or current periods, the quantities are fixed at some other period b:

$$P_{\text{Lowe}} = \frac{\sum_i p_i^t q_i^b}{\sum_i p_i^0 q_i^b} \tag{5.15}$$

The Lowe Index is particularly important as it is the formula that almost all National Statistical Institutes use for their consumer price indices, with the period b representing some period *earlier* than the base period to ensure that reliable weighting information can be obtained. In much of the literature, the Lowe index is described as a 'Laspeyres-type' index; however, one could argue that a more accurate description is that the Laspeyres is a special case of the Lowe index where period b coincides with the base period. Similarly, the Paasche price index is a special case of the Lowe index where period b coincides with the current period.

A slightly different formula incorporating a weight reference period b is the **Young price index**:

$$P_{\text{Young}} = \sum_i \left(\frac{p_i^t}{p_i^0}\right).s_i^b, \quad \text{with } s_i^b = \frac{p_i^b q_i^b}{\sum_j p_j^b q_j^b} \tag{5.16}$$

This has the weights as period b expenditure shares.

5.8   Chapter summary

In this chapter, we have introduced two very well-known price and quantity index formulae and have seen how they incorporate price and quantity information in different ways. Both measure the overall price change by fixing quantities and measure the quantity change by fixing prices, but the period at which those prices and quantities are fixed differs. These differences lead to differing measures of price and quantity change.[3]

In practice, a Laspeyres price index is preferred to the Paasche equivalent due to the advantage of only having to collect one set of quantities. However, the actual formula used by almost all National Statistics Institutes is the Lowe index where the period from which weighting information is taken precedes the base period.

In addition, we have seen that Laspeyres and Paasche indices can be combined to provide a solution to the Index Number Problem.

References

1. See Walsh, C.M. (1901) *The Measurement of General Exchange-Value*, The Macmillan Company, New York, p. 425.
2. See ILO (2004) *Consumer Price Index Manual: Theory and Practice*, International Labour Office, Geneva.

---

[3] Laspeyres and Paasche indices can produce the same index numbers. In the case of price indices, the quantities would need to be the same for both time periods and vice versa for quantity indices.

Exercise D

A full set of solutions to this exercise can be found in Appendix D.

D.1   Table D.1 presents the prices of and quantities of three items sold by the menswear section of a department store in two consecutive months.

Table D.1   Price and Quantity of clothing June to July.

		Price (£)		Quantity sold	
		June	July	June	July
1	Mens formal shirt	35.00	37.50	200	100
2	Mens formal trousers	30.00	25.00	250	300
3	Tie	18.00	18.00	50	45

Using June as the base period, calculate the following for July:

a. The Laspeyres price index number.

b. The Paasche price index number.

c. The Laspeyres quantity index number.

d. The Paasche quantity index number.

D.2   Table D.2 presents the quantity of meat sold by a large supermarket in two separate months in 2014 alongside the corresponding price data.
  Using June 2014 as the base period, calculate the following for December 2014:

a. The Laspeyres price index number.

b. The Paasche price index number.

c. The Laspeyres quantity index number.

d. The Paasche quantity index number.

e. Comment on your answers to parts a–d.

Table D.2   Price and quantity for meat June and December.

	Price (£)		Quantity (thousands)	
	June 2014	December 2014	June 2014	December 2014
Turkey	35.00	50.00	2	50
Ham	15.00	20.00	13	28
Chicken	7.00	7.00	40	18
Lamb	22.00	18.00	32	3

D.3 Table D.3 presents the prices relatives between two periods 0 and i for five items sold by a pharmacy along with their value share in the base period.

Table D.3   Price and value data for a pharmacy.

	$R^{0,t}$	s^0
Toothpaste	1.34	0.20
Conditioner	1.22	0.10
Shampoo	1.32	0.35
Dental floss	1.01	0.05
Deodorant	1.12	0.30

Calculate the Laspeyres price index number for these items in period t with period 0 as the base period.

D.4 Consider an application of index numbers to poultry farming.

In any particular month, the number of eggs laid per chicken can be thought of as analogous to prices for economic index numbers. Similarly, the number of chickens is analogous to quantity and the number of eggs is analogous to value.

Using data from Table D.4 relating to farms in a small town, calculate the following:

a. The Laspeyres egg-rate (number of eggs laid per chicken) index number in April, using March as the base period.

b. The Paasche egg-rate index number in April, using March as the base period.

c. What would an index for number of chickens quantify? Calculate the Laspeyres index number for the number of chickens.

Table D.4   Chicken and egg data, March and April.

	Number of chickens		Number of eggs laid	
	March	April	March	April
Farm A	23	28	698	715
Farm B	31	25	851	864
Farm C	84	92	2222	2539
Farm D	58	61	1134	1211
Farm E	21	22	603	628

6

Domains and aggregation

The main measure of the UK consumer price inflation, the Consumer Prices Index, was briefly described in Chapter 1. The 'all-items', 12-month growth measure is the headline inflation figure in news bulletins; for example, the annual rate of inflation in the UK, as measured by the 12-month growth in the Consumer Prices Index (CPI), for May 2014 was 1.5% (down from 1.8% in April).[1] This chapter considers values of inflation for specific categories of goods and services and how they are related to the all-items figure.

When measuring price changes for all consumer goods and services on offer in the UK, which included goods as diverse as cars, sliced bread, washing machines and haircuts, one number can only explain so much. This overall grouping of commodities can be usefully subdivided into smaller collections of goods and services, or 'domains'. If these domains are defined below the headline measure in a meaningful manner, then calculating the index numbers for each domain will help us better understanding the overall figure.

Returning to the example of the UK CPI in May 2014; the 12-month growth in the indices for a selection of domains is presented in Table 6.1. We can see from this table that while prices in the UK economy as a whole increased by 1.5% in the year to May 2014, the price of the 'Alcoholic beverages and tobacco' domain increased at a much higher rate (4.6%) and the general level of prices for 'Food and non-alcoholic beverages' fell.

6.1 Defining domains

In order for an index number calculated for a domain to have practical value, the domain for which it is calculated needs to be carefully defined. It must make sense to group items together and there should be an interest in the measurement of the overall

---

[1] See http://www.ons.gov.uk/ons/rel/cpi/consumer-price-indices/may-2014/stb---consumer-price-indi ces---may-2014.html

A Practical Introduction to Index Numbers, First Edition. Jeff Ralph, Rob O'Neill and Joe Winton.
© 2015 John Wiley & Sons, Ltd. Published 2015 by John Wiley & Sons, Ltd.
Companion Website: http://www.wiley.com/go/ralph/index_numbers

Table 6.1   UK consumer prices index, percentage change
on a year earlier selected CPI divisions, May 2014.

	Percentage change on a year earlier, May 2014
Food and non-alcoholic beverages	−0.6
Alcoholic beverages and tobacco	4.6
Transport	0.4
Recreation and culture	1.1
CPI (overall index)	1.5

Source: Office for National Statistics.

change in price or quantity for that group. One way to achieve this is by grouping 'similar' goods and services together. If goods in the same domain are similar to each other, then it is likely that we will want to know about the price and quantity movements of the whole group and that these prices and quantities will move in a similar way. For example, it is likely that the price movements of garden spades and forks will be similar, though it is not necessarily the case.

If the overall set of goods and services has been split into a number of discrete domains, then it is also possible to split each of those domains further into a number of sub-domains. In this way, an 'aggregation structure' comprising many categories and many levels can be constructed; index numbers can then be calculated at each level.

An example of such a structure is COICOP – Classification of Individual Consumption by Purpose, which is defined by the Statistics Division of the United Nations[2] and divides individual consumption into a large number of categories. Eurostat uses COICOP to specify the levels at which European countries should measure the change in the general level of prices as part of the main European Harmonised Index of Consumer Prices (HICP)[3]; in the UK, this is the CPI.

Figure 6.1 gives an example of how COICOP is used in the UK to define domains within the CPI. The overall range of available goods and services is split into a number of broad groups such as 'Transport', 'Clothing and footwear' and 'Alcohol and Tobacco'. Each of these is then split into more detailed domains; for example, the domain 'Alcohol and Tobacco' is further split into two sub-domains, 'Alcohol' and 'Tobacco'. These can each be split again to create more specific groups.

Another important factor in defining domains is comparability. As we saw in Chapter 2, one of the benefits of using index numbers is to enable us to make useful comparisons; we should, therefore, try to make our domains as comparable

---

[2] See COICOP http://unstats.un.org/unsd/cr/registry/regcst.asp?Cl=5, Uses of COICOP in the EU http://unstats.un.org/unsd/class/intercop/expertgroup/2007/AC124-27.PDF

[3] This index is described in Section 11.2.

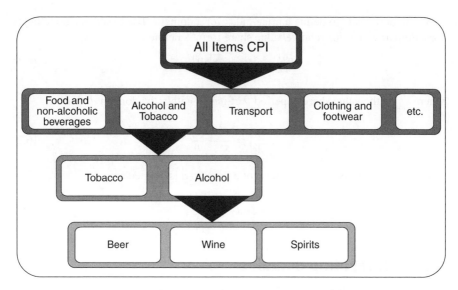

Figure 6.1   Example of COICOP structure.

as possible with other countries, with other time periods and with other measures. COICOP is a good example of this; the structure has been adopted by all member states of the EU for calculating a consistent measure of consumer price inflation. As a result, meaningful comparisons of the consumer price movements between countries can be made. For example, we can compare how food costs have changed in Germany relative to the UK. Because the COICOP structure is largely fixed across time, it means that comparisons of index numbers over time can also be made.

6.2   Indices for domains

To calculate index numbers for domains, we first split all of the goods and services in the basket into K discrete domains. Domain index numbers can then be calculated for each domain $d = 1, \dots, K$ by treating each domain as a separate collection of goods and services and using the index number formulae discussed in Chapters 4 and 5.

Before giving the formula for a domain index, we need to extend the mathematical notation, which we have been using up to now. It is convenient to consider the items in a domain as members of a set and use the basic notation of mathematical sets. So, if the domain is 'food', then instead of adding up the price changes for all items in the basket, we count only food items. We say that the food domain is a set, and we use the following mathematical set notation: $i \in d$, to mean 'just include the values of i that correspond to the food items in the food domain d'. Our sum now looks like this: $\sum_{i \in d} x_i$.

We are now ready to define a price index for a domain. The Laspeyres price index for a domain d in price relative/value share form is defined as:

$$P^{0,t}_{\text{Laspeyres};\,d} = \sum_{i \in d} R^{0,t}_i \cdot s^0_{i;d} \tag{6.1}$$

where $s^0_{i;d}$ is the value share at period 0 of item i within the domain d, that is:

$$s^0_{i;d} = \frac{v^0_i}{\sum_{i \in d} v^0_i} \tag{6.2}$$

Similarly, the Paasche price index for a domain d is defined as:

$$P^{0,t}_{\text{Paasche};\,d} = \frac{1}{\sum_{i \in d} \frac{s^t_{i;d}}{R^{0,t}_i}} \tag{6.3}$$

6.3   Aggregating domains

Once indices for each domain have been specified, we can then create indices for the aggregated domains. In some instances, it is possible to combine the domain indices to calculate a higher level index, which takes the same value as if we had calculated the higher level index directly from the price and quantity data. Where this is possible, we say that the index is **'consistent in aggregation'**. This simplifies the construction of indices at multiple levels of a classification scheme.

> An index is said to be consistent in aggregation when the index for some aggregate has the same value whether it is calculated directly in a single operation, without distinguishing its components, or it is calculated in two or more steps by first calculating separate indices or sub-indices, for its components or subcomponents, and then aggregating them, the same formula being used at each step.
>
> OECD Definition[4]

Consistency in aggregation is dependent on the mathematical form of an index. Both the Laspeyres and Paasche price indices are consistent in aggregation. This means that for the overall collection of all goods and services D, split into K domains, $d = 1, \ldots, K$, a Laspeyres price index for D can be calculated as:

$$P^{0,t}_{\text{Laspeyres};\,D} = \sum_{d \in D} P^{0,t}_{\text{Laspeyres};\,d} \cdot s^0_d \tag{6.4}$$

where s^t_d is the value share of domain d within the total basket D at period t:

$$s^0_d = \frac{\sum_{i \in d} v^0_i}{v^0} \tag{6.5}$$

And $P^{0,t}_{\text{Laspeyres};\,d}$ is the Laspeyres price index for domain d.

---

[4] OECD Glossary of Statistical Terms, http://stats.oecd.org/glossary/detail.asp?ID=5608

Similarly, the Paasche price index for the total basket D, split into K domains, $d = 1, \dots, K$, can be calculated as:

$$P^{0,t}_{\text{Paasche};D} = \cfrac{1}{\sum_{d \in D} \cfrac{s^t_d}{P^{0,t}_{\text{Paasche};d}}} \tag{6.6}$$

Example 6.1   Consider the sales data given in Table 6.2. The owner of *Joe's Bike Shop* wants to use a Laspeyres Price Index to measure the change in price of the items that are sold in the shop. The products sold can be easily grouped into three distinct categories: accessories, bikes and clothing.

a) Calculate the Laspeyres price index numbers for each of the three categories in 2010 with 2005 as the base period

b) Calculate the Laspeyres price index number for the whole shop in 2010 with 2005 as the base period by combining your answers to a)

c) Calculate the Laspeyres price index number for the whole shop in 2010 with 2005 as the base period directly from the price and value data.

Table 6.2   Sales information from Joe's bike shop 2005–2010.

	Pounds (£)		
	2005 turnover (v^0)	2005 price (p^0)	2010 price (p^1)
Accessories			
Water bottle	960	8	10
Puncture repair kit	750	3	3
Total accessories	*1710*		
Bikes			
Mens road bike	16 500	300	350
Mens mountain bike	6750	225	230
Ladies bike	3750	250	200
Childs bike	8400	120	110
Total bikes	*35 400*		
Clothing			
Helmet	5500	22	25
Cycling jersey	2900	29	35
Cycling shoes	3230	38	42
Total clothing	*11 630*		

Solution

a) First, we need to calculate the turnover share of each item within each category, and the price relative for 2010 with 2005 as the base period as follows (Table 6.3):

Table 6.3   Joe's bike shop turnover shares and price relatives.

	2005 turnover (v^{2005})	2005 turnover share within domain $(s^{2005}_{i;Domain})$	Price relative 2010 $(R^{2005,2010})$
Accessories			
Water bottle	960	$960/1710 = 0.561$	$10/8 = 1.250$
Puncture repair kit	750	$750/1710 = 0.439$	$3/3 = 1.000$
Total accessories	*1710*		
Bikes			
Mens road bike	16 500	$16\,500/35\,400 = 0.466$	$350/300 = 1.167$
Mens mountain bike	6750	$6750/35\,400 = 0.191$	$230/225 = 1.022$
Ladies bike	3750	$3750/35\,400 = 0.106$	$200/250 = 0.800$
Childs bike	8400	$8400/35\,400 = 0.237$	$110/120 = 0.917$
Total bikes	*35 400*		
Clothing			
Helmet	5500	$5500/11\,630 = 0.473$	$25/22 = 1.136$
Cycling jersey	2900	$2900/11\,630 = 0.249$	$35/29 = 1.207$
Cycling shoes	3230	$3230/11\,630 = 0.278$	$42/38 = 1.105$
Total clothing	*11 630*		

We can then calculate the three Laspeyres price index numbers as:

$$P^{2005,2010}_{Laspeyres;Accessories} = 100 \times \sum_{i=1}^{2} R_i^{2005,2010} \cdot s^{2005}_{i;Accessories}$$

$$= 100 \times ((1.250 \cdot 0.561) + (1.000 \cdot 0.439)) = 114.0$$

$$P^{2005,2010}_{Laspeyres;Bikes} = 100 \times \sum_{i=1}^{4} R_i^{2005,2010} \cdot s^{2005}_{i;Bikes}$$

$$= 100 \times ((1.167 \cdot 0.466) + (1.022 \cdot 0.191)$$

$$+ (0.800 \cdot 0.106) + (0.917 \cdot 0.237)) = 104.1$$

$$P^{2005,2010}_{Laspeyres;Clothing} = 100 \times \sum_{i=1}^{3} R_i^{2005,2010} \cdot s^{2005}_{i;Clothing}$$

$$= 100 \times ((1.136 \cdot 0.473) + (1.207 \cdot 0.249)$$

$$+ (1.105 \cdot 0.278)) = 114.5$$

b) Next, we need to calculate the turnover share of each category in the shop (Table 6.4):

Table 6.4   Joe's bike shop turnover and turnover shares.

Domain	2005 turnover (v^{2005})	2005 domain turnover share domain (s^{2005}_{Domain})
Accessories	1710	$1710/48\,740 = 0.035$
Bikes	35 400	$35\,400/48\,740 = 0.726$
Clothing	11 630	$11\,630/48\,740 = 0.239$
Total	48 740	

We can then calculate the Laspeyres price index number as:

$$P^{2005,2010}_{Laspeyres} = \sum_{d=1}^{D} P^{2005,2010}_{Laspeyres;d} \cdot s^{2005}_d$$

$$= (P^{2005,2010}_{Laspeyres;Accessories} \cdot s^{2005}_{Accessories}) + (P^{2005,2010}_{Laspeyres;Bikes} \cdot s^{2005}_{Bikes})$$

$$+ (P^{2005,2010}_{Laspeyres;Clothing} \cdot s^{2005}_{Clothing})$$

$$= (114.0 \cdot 0.035) + (104.1 \cdot 0.726) + (114.5 \cdot 0.239) = 106.9$$

c) To calculate the Laspeyres price index number directly from the turnover and price index, we need to calculate the turnover share of each item within the shop:

	2005 turnover, v^{2005}	2005 turnover share, s^{2005}_i
Water bottle	960	$960/48\,740 = 0.020$
Puncture repair kit	750	$750/48\,740 = 0.015$
Mens road bike	16 500	$16\,500/48\,740 = 0.339$
Mens mountain bike	6750	$6750/48\,740 = 0.138$
Ladies bike	3750	$3750/48\,740 = 0.077$
Childs bike	8400	$8400/48\,740 = 0.172$
Helmet	5500	$5500/48\,740 = 0.113$
Cycling jersey	2900	$2900/48\,740 = 0.059$
Cycling shoes	3230	$3230/48\,740 = 0.066$
Total	48 740	

We can then calculate the Laspeyres price index number as usual:

$$P^{2005,2010}_{Laspeyres} = 100 \times \sum_{i=1}^{9} R^{2005,2010}_i \cdot s^{2005}_i$$

$$= 100 \times \begin{pmatrix} (1.250 \cdot 0.020) + (1.000 \cdot 0.015) + (1.167 \cdot 0.339) \\ + (1.022 \cdot 0.138) + \\ (0.800 \cdot 0.077) + (0.917 \cdot 0.172) + (1.136 \cdot 0.113) \\ + (1.207 \cdot 0.059) + (1.105 \cdot 0.066) \end{pmatrix}$$

$$= 106.9$$

6.4   More complex aggregation structures

We have just seen how to calculate Laspeyres and Paasche index numbers for domains of the total basket and how these can be combined to calculate Laspeyres and Paasche index numbers for the overall collection of goods and services, but what if we have a more detailed aggregation structure, that is, for sub-domains as well as domains?

To calculate an index number for a sub-domain, we work in exactly the same way as in the rest of this chapter, but the group of items over which we calculate our index numbers just gets smaller. Calculating indices for sub-domains within domains is the same process as calculating indices for domains.

6.5   A note on aggregation structures in practice

We have seen here that it is possible to calculate indices for different domains within an overall set of goods and services and, if we choose an appropriate formula, that these can be combined to calculate higher level index numbers.

At each level of our aggregation structure, we have used weights to combine price movements to calculate higher level index numbers. At the bottom of the structure, we often do not have turnover information, so this is not possible. For example, we can work out the turnover share of orange juice compared with cola, but within the cola domain, it is very difficult to work out the turnover shares for each brand of cola. In this case, we will need to combine the prices using an un-weighted formula, as we did in Chapter 4. As a result, we would like to get to a point where it seems reasonable to assume that there is not too much difference between items (or their price behaviour) before relying on an un-weighted index. This problem will be discussed in more detail in Chapter 14.

If our classification scheme has been defined well, then at the bottom of the structure, the products will be relatively 'homogeneous' in terms of price behaviours – this means that prices move in a similar way within the low-level sub-domain. This is a very important feature of an aggregation structure, which in practice is difficult to achieve. Consider, for example a price index of chocolate bars. Some consumers might argue that all chocolate bars of approximately the same size are the same; however, a confectionary aficionado might argue that they perceive these goods as being very different. We might then need to consider whether treating chocolate as being homogeneous is reasonable. We might consider at least one further subdivision, perhaps into standard and luxury chocolate bars.

6.6   Non-consistency in aggregation

Combining domain index numbers to calculate index numbers at higher levels of aggregation works well for the Laspeyres and Paasche indices; however, it does not work for all index formulae. We will introduce geometric index formulae in later chapters where consistency in aggregation does not apply. This is not a serious issue; however, it is more of an inconvenience. It means that for non-consistent formulae, we have to use the price relatives and weights for the whole domain each time we calculate a domain index.

6.7   Chapter summary

In this chapter, we have seen that it is possible to calculate index numbers for different domains within an overall collection of goods and services and then to combine the index numbers for sub-domains and domains at each level of the aggregation structure to end up with an overall index number.

Index numbers for different domains can provide invaluable insight into different sectors of the economy and help us gain a better understanding of the changes in a headline figure. This chapter has shown that the index formulae that we have learned so far in this book can also be applied to domains of goods and services.

Exercise E

Solutions to this exercise can be found in Appendix D.

E.1   Table E.1 presents price relatives of 11 products sold by a bar between 2 years, 2013 and 2014. These products can be split into three categories: Beer, Wine and Spirits and the table also gives the value share in 2013 of each product within these categories.

Table E.1   Price and value data for a bar, 2013–2014.

		Category value share $s_{i;d}^{2013}$	Price relative $R_i^{2013,2014}$
Beer	Lager (pint)	0.45	1.25
	Lager (bottle)	0.20	1.30
	Bitter (pint)	0.35	1.20
Wine	New World red	0.30	1.01
	New World white	0.15	1.10
	European red	0.20	1.05
	European white	0.20	1.00
	Sparkling	0.15	1.12
Spirits	Vodka	0.60	0.95
	Whisky	0.30	0.97
	Gin	0.10	0.92

a. Calculate the Laspeyres price index numbers for each of the three categories: Beer, Wine and Spirits in 2014 with 2013 as the base period.
   You are now given value shares for each category in 2013 (Table E.2).

Table E.2   Value shares for drink categories.

	Overall value share
Beer	0.50
Wine	0.30
Spirits	0.20

b. Calculate the Laspeyres price index number for the whole pub in 2014 with 2013 as the base period.

E.2   Table E.3 shows price and quantity information for a selection of newspapers and magazines sold by a newsagent in two months of the same year, January and May. These data can be easily divided into two domains, 'Newspapers' and 'Magazines'

Table E.3   Price and quantity data for newspapers and magazines.

		Price (£)		Quantity	
		January	May	January	May
Newspapers	Broadsheet	0.90	1.00	150	145
	Tabloid	0.35	0.35	300	350
	Local	0.40	0.65	50	26
Magazine	Lifestyle	1.50	1.45	250	300
	Fashion	2.00	2.20	500	420
	Fitness	3.00	3.50	120	100
	Home and garden	4.00	4.00	120	160

a. Calculate the Fisher price index number for the category 'Newspapers'.

b. Calculate the Fisher price index number for the category 'Magazines'.

c. Calculate the Fisher price index number for the whole shop.

E.3   Listed in Table E.4 are 10 items sold by a gift shop. The owner of the shop wishes to calculate the price index numbers for different categories of goods sold by the shop. Devise a classification system for these goods to enable the owner to do this.

You can include as many categories as you wish, and price relatives for these items are also included to help your decisions.

Explain your decisions.

Table E.4   Items sold by a gift shop and example price relatives.

	$R^{0,t}$
Coaster	1.94
Framed photo	0.88
Guidebook	1.92
Key ring	1.89
Mug	1.05
Photograph print	0.89
Post card	1.91
Spoon set	1.13
Sweets	1.94
Tea pot	1.06
Tea set	1.15
Watercolour print	0.82

7

Linking and chain-linking

In previous chapters, we have seen how index numbers can be used to make comparisons across time. In order for these comparisons to be most effective, we aim for our index numbers to be constructed on the same bases at every point in the series. However, in practice, this ideal is difficult to maintain.

A competing principle is that when calculating index numbers, we aim to make the index representative of the circumstances that we are measuring. In the case of the Consumer Prices Index, we need to reflect the goods and services currently available to consumers and in Chapter 1, we described the use of a *representative* basket of goods of services for which we measure prices. Changes in the buying habits of consumers will affect how representative our basket is, and therefore, it is important to periodically revisit and refresh the basket to ensure that it continues to be relevant. We also need to consider new products and new technology; for example, if we construct a sub-index for the domain 'books', we need to think about how the index will be affected by the rise in popularity of e-readers.

There are other sources of change. In Chapter 6, we introduced the idea of a classification structure as a way of organising goods and services to provide a useful and consistent grouping for reporting and comparison purposes. Occasionally, there are changes in this classification structure and these may have an effect on an index number series.

These two principles can be summarised as follows: there is value in a long run series calculated on a consistent basis, which does not accommodate change; but by achieving this, the series gradually becomes less representative and less relevant.

If we do decide to change the way that our index number series is constructed, then we are faced with the likelihood of a discontinuity in the series at the point of change. Such a 'structural' change is not the same as a change in prices but rather an effect of a change in the method of calculation.

In order to get the most out of an index series, we aim for a series that is continuous but also maintains relevance to current conditions.

A Practical Introduction to Index Numbers, First Edition. Jeff Ralph, Rob O'Neill and Joe Winton.
© 2015 John Wiley & Sons, Ltd. Published 2015 by John Wiley & Sons, Ltd.
Companion Website: http://www.wiley.com/go/ralph/index_numbers

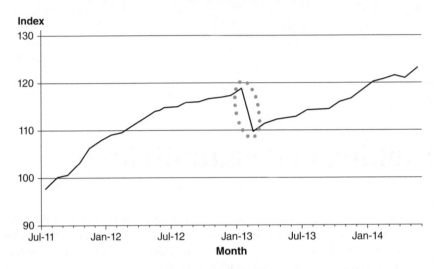

Figure 7.1   Price index of white goods, July 2011–June 2014.

Consider the chart in Figure 7.1. This shows a fictional price index number series for white goods (for example, washing machines and fridges) over a 3-year period. In January 2013, there was a change to the classification system used to group items together. The graph of the index series shows a large fall in January 2013, but beyond this point, the series continued to rise as before. This discontinuity or step change in the index series is unlikely to be caused by a real change in the price of white goods but instead is likely to be caused (at least in part) by the change in the construction of the index series.

Given that the fall in the series at January 2013 is unlikely to be an actual price change, it might be clearer to present the series with a break as in Figure 7.2. In Figure 7.2, it is clear that the two series are different but it is now difficult to make meaningful comparisons across time: how can we compare price index numbers for white goods in January 2012 and January 2014?

For many users of index numbers, it is the ability to make such meaningful comparisons that is important; we therefore need a way to eliminate this discontinuity. To do this, we can either adjust the series calculated on the new basis to continue the original series or adjust the original series so that it continues back from the series calculated on the new basis as shown in Figure 7.3.

7.1   Linking

The solution to this problem is **linking.** To link two series together, we need to choose a **link period.** In the link period, index numbers are calculated on both the old and new bases; the growth in the series in this link period on the old basis is then applied to the rest of the index number series calculated on the new basis to create a linked series.

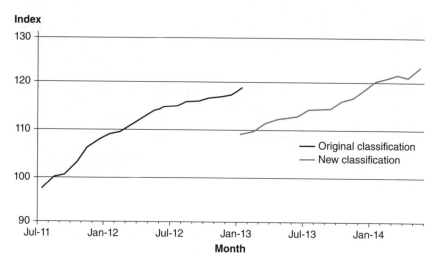

Figure 7.2   Price index of white goods under old and new classification systems, July 2011–June 2014.

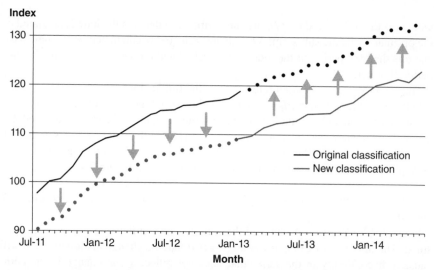

Figure 7.3   Options for linking price indices of white goods under old and new classification systems, July 2011–June 2014.

The usual practice is to leave the index series unchanged up to the link period and adjust the series from that point on to avoid continual revisions to past data. A linked index number series is then calculated as:

$$
I_{LINKED}^{0,t} = \begin{cases} I_{OLD}^{0,t} & t \leq l \\ \dfrac{I_{OLD}^{0,l}}{I_{NEW}^{a,l}} I_{NEW}^{a,t} & t > l \end{cases}
\tag{7.1}
$$

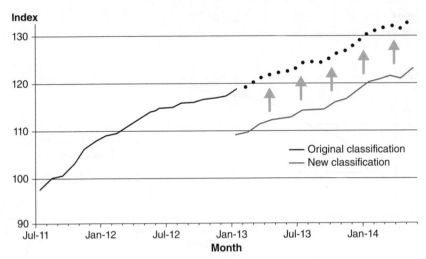

Figure 7.4   Linking price indices of white goods under old and new classification systems, July 2011–June 2014.

where l is the link period $(0 < l)$, the sub-script *OLD* denotes the index calculated on the original basis, the sub-script *NEW* denotes the index calculated on the new basis and a is the base period of the index series calculated on the new basis; this is often period 0 or period l.

The ratio $I_{OLD}^{0,l}/I_{NEW}^{a,l}$ is often referred to as the link factor.

In the example shown in Figures 7.1–7.3, this would mean selecting January 2013 as the link period and calculating price index numbers for this period under both the original and the new classification systems. The ratio of the index numbers in the link period under each classification can then be calculated, and this link factor can then be applied to the index series beyond the link period as in Figure 7.4.

Example 7.1 Table 7.1 shows an index of quarterly turnover for Company A in 2012 and 2013.

In Q1 2013, Company A changed the way that certain items of turnover are classified. The directors of Company A are worried that the change in classification will cause a discontinuity in the series that does not reflect a real change in the company's turnover. The index number has been calculated for Q1 2013 under the old classifications system to be 118.

Table 7.1   Quarterly index of turnover for company A, 2012–2013.

Index (Q1 2012 = 100)							
Q1 2012	Q2 2012	Q3 2012	Q4 2012	Q1 2013	Q2 2013	Q3 2013	Q4 2013
100	115	110	120	85	92	89	100

Calculate a linked series for the rest of 2013, using Q1 2013 as the link period.
Solution Calculate the link factor as:

$$\frac{I^{0,l}_{OLD}}{I^{a,l}_{NEW}} = \frac{I^{0,Q1\,2013}_{OLD}}{I^{0,Q1\,2013}_{NEW}} = \frac{118}{85} = 1.388\ldots$$

Then, for Q2 2013:

$$I^{0,Q2\,2013}_{LINKED} = \frac{I^{0,Q1\,2013}_{OLD}}{I^{0,Q1\,2013}_{NEW}} I^{0,Q2\,2013}_{NEW} = 1.388 \times 92 = 127.7$$

and for Q3 2013:

$$I^{0,Q3\,3013}_{LINKED} = \frac{I^{0,Q1\,2013}_{OLD}}{I^{0,Q1\,2013}_{NEW}} I^{0,Q3\,2013}_{NEW} = 1.388 \times 89 = 123.6$$

and for Q2 2013

$$I^{0,Q4\,2013}_{LINKED} = \frac{I^{0,Q1\,2013}_{OLD}}{I^{0,Q1\,2013}_{NEW}} I^{0,Q4\,2013}_{NEW} = 1.388 \times 100 = 138.8$$

7.2 Re-basing

A primary reason for linking is to allow for the update of the weights used in the
construction of the index. Let us take an example of a price index for alcoholic drinks
sold in a particular pub where the prices of lager and cider are measured each quarter
and then combined to form an aggregate index using the turnover share of the two
items as weights. Suppose that traditionally lager was more popular than cider as
shown in Table 7.2, but in 2011, cider became much more popular.

The weights used to calculate the combined index numbers will have a big effect
on its level as shown in Figure 7.5. It is reasonable to adjust the weights in 2011 to
better represent the buying habits of the pub's customers, but changing the weights
from Q1 2011 onwards would cause a big step change in the index numbers series.

Linking can be used to update the base period and therefore the weights; this is
especially important when updating the basket.

Table 7.2   Turnover share for Lager and Cider.

	2010 weight	2011 weight
Lager	0.9	0.3
Cider	0.1	0.7

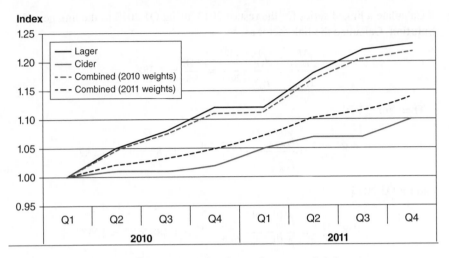

Figure 7.5   Combined alcohol index under two weighting patterns.

Example 7.2 Consider a price index of personal grooming services. In period 0, a basket of three items is chosen. This basket is fixed and a Laspeyres price index is used to calculate the overall index numbers (Table 7.3).

$$P^{0,t}_{\text{Laspeyres}} = 100 \times \sum_{i=1}^{3} R_i^{0,t} \times s_i^0$$

$$P^{0,0}_{\text{Laspeyres}} = 100$$

$$P^{0,1}_{\text{Laspeyres}} = 100 \times \sum_{i=1}^{3} R_i^{0,1} \times s_i^0 = 100$$

$$\times ((1.05 \times 0.2) + (1.06 \times 0.6) + (1.02 \times 0.2)) = 105$$

Table 7.3   Data for personal grooming services – 1.

	$R^{0,0}$	$R^{0,1}$	s^0
Wet shave	1.00	1.05	0.2
Hair cut	1.00	1.06	0.6
Manicure	1.00	1.02	0.2

In period $t = 2$, a decision is made to update the basket. Due to changes in consumer behaviour, 'spray tan' is added to the basket and 'wet shave' is removed. As a result, value shares are now calculated for period $t = 2$ and for periods following $t = 2$; it is now not possible to calculate the index numbers with $t=0$ as the base period as we do not have the $t=0$ expenditure share for the new item.

Instead, we calculate new Laspeyres price index numbers with $t=2$ as the base period (Table 7.4).

$$P^{2,2}_{\text{Laspeyres}} = 100$$

$$P^{2,3}_{\text{Laspeyres}} = 100 \times \sum_{i=1}^{3} R^{2,3}_i \times s^2_i = 100$$

$$\times ((1.07 \times 0.5) + (1.04 \times 0.1) + (1.11 \times 0.4)) = 108.3$$

$$P^{2,4}_{\text{Laspeyres}} = 100 \times \sum_{i=1}^{3} R^{2,4}_i \times s^2_i = 100$$

$$\times ((1.06 \times 0.5) + (1.05 \times 0.1) + (1.10 \times 0.4)) = 107.5$$

Table 7.4   Data for personal grooming services – 2.

	$R^{2,2}$	$R^{2,3}$	$R^{2,4}$	s^2
Wet shave	–	–	–	0
Hair cut	1.00	1.07	1.06	0.5
Manicure	1.00	1.04	1.05	0.1
Spray tan	1.00	1.11	1.10	0.4

In order to construct a continuous series from $t=0$ to $t=4$, we need to link the two series together. If we choose $t=2$ as the link period, then we need to calculate the index numbers in period $t=2$ with $t=0$ as the base period (Table 7.5).

$$P^{0,2}_{\text{Laspeyres}} = 100 \times \sum_{i=1}^{3} R^{0,2}_i \times s^0_i = 100$$

$$\times ((1.07 \times 0.2) + (1.06 \times 0.6) + (1.05 \times 0.2)) = 106$$

Table 7.5   Data for personal grooming services – 3.

	$R^{0,2}$	s^0
Wet shave	1.07	0.2
Hair cut	1.06	0.6
Manicure	1.05	0.2
Spray tan	–	0

We can then calculate the linked series:

$$P^{0,t}_{\text{Laspeyres,LINKED}} = \begin{cases} P^{0,t}_{\text{Laspeyres}} & t \le 2 \\[2ex] \dfrac{P^{0,2}_{\text{Laspeyres}}}{P^{2,2}_{\text{Laspeyres}}} \times P^{2,t}_{\text{Laspeyres}} & t > 2 \end{cases}$$

$$P^{0,3}_{\text{Laspeyres,LINKED}} = \frac{P^{0,2}_{\text{Laspeyres}}}{P^{2,2}_{\text{Laspeyres}}} \times P^{2,3}_{\text{Laspeyres}} = \frac{106}{100} \times 108.3 = 114.8$$

$$P^{0,4}_{\text{Laspeyres,LINKED}} = \frac{P^{0,2}_{\text{Laspeyres}}}{P^{2,2}_{\text{Laspeyres}}} \times P^{2,4}_{\text{Laspeyres}} = \frac{106}{100} \times 107.5 = 114.0$$

7.3   Chain-linking

Chain-linking is a term given to the regular-usually annual-linking of an index series, commonly as the result of updated weights and sample changes.

In Figure 7.6, an index series has been calculated for the 3 years, from 2008 to 2010, with January in each year as the base period. This situation is found where a basket of items and weights are updated every January. In this situation, meaningful comparison across time is difficult unless it is being made within the same year.

In Figure 7.7, the series have been linked together using January 2009 and January 2010 as link periods. With the linked series shown in Figure 7.7, we can now make comparisons across time using the most up-to-date items and weights to construct the index in each year.

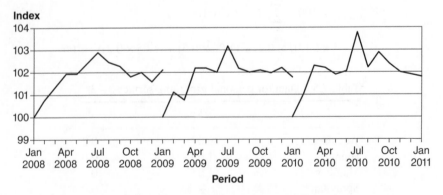

Figure 7.6   Index series with January base each year, 2008–2010.

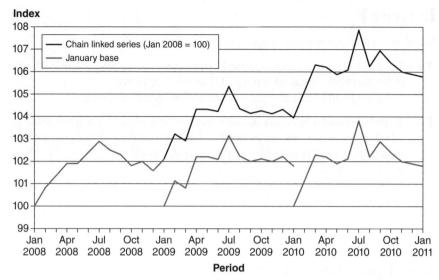

*Figure 7.7   Index series with January base each year and annually chain-linked,
2008–2010.*

7.4   Chapter summary

In previous chapters, we have seen the practical advantage of calculating
Laspeyres-type, base-weighted price index numbers in terms of cost and time-
liness. The risk with fixing a basket and the weights of each item at their base period
values is that over time they become unrepresentative and the index series becomes
less valuable in interpreting movements in the prices of current goods and services.

We have seen in this chapter how this problem can be overcome. By linking
together index series with an updated basket and weights, we can update our index
in response to the changing conditions of the marketplace while maintaining a continuous series.[1] Linking can also be used to accommodate changes in methodology
where, for example, an improved aspect of a calculation has been developed.

---

[1] However, annual linking does not reflect changes within a calendar year.

Exercise F

Solutions to this exercise can be found in Appendix D.

F.1 Table F.1 shows the same quarterly index series over 2 years; in each year, index numbers are calculated with Q1 as the base period.
   Using Q1 2013 as a link period, link the two series together.

F.2 Table F.2 shows a selection of index numbers calculated for a set of data.
   In each period, t, the index number has been calculated with the previous month $(t-1)$ as the base period. For example, in period 4, $I^{(t-1,t)} = I^{(3,4)}$ and gives the change between period 3 and period 4.
   Calculate a linked index series for periods 1–7. Use each period $t-1$ as a link period where $I^{(t-1,t-1)} = 100$.

Table F.1   Index series, 2012–2013.

	2012				2013			
	$q1$	$q2$	$q3$	$q4$	$q1$	$q2$	$q3$	$q4$
Old	100.00	101.30	100.80	102.60	103.10			
New					100.00	101.60	100.50	102.90

Table F.2   Index numbers with previous period base.

t	$I^{(t-1,t)}$
1	100.0
2	100.6
3	103.0
4	101.3
5	103.7
6	101.3
7	100.7

F.3 Table F.3 presents index series for the 3 years, from 2001 to −2003. In each year, index numbers are constructed with quarter one (Q1) as the base period.

Table F.3   Index series, 2001–2013.

	Index (Q1 Y base)		
	2001	2002	2003
Q1 Y	100	100	100
Q2 Y	103	104	105
Q3 Y	104	104	103
Q4 Y	103	104	105
Q1 Y + 1	108	106	

a. Using Q1 2002 as a link period, link the series for 2001 and 2002 together.

b. Using Q1 2003 as a link period, link the series for 2003 onto the linked series calculated in a).

8

Constructing the consumer prices index

In Chapters 4 and 5, we saw how price and weighting data can be used to estimate the overall price change for a collection of goods and services. Chapters 6 and 7 then introduced two important practical aspects of index number construction – aggregation and chain-linking. Having covered this material, we are now in a position to explore how consumer price inflation is measured in practice through the Consumer Prices Index (CPI). Although the description of the index in this chapter concerns the UK CPI, the topics covered apply almost universally to how the equivalent measure is put together in other countries.

The full specification of the CPI would take a whole book in itself; this chapter therefore provides just an overview. Other reference documents provide a much fuller description. These include the technical specification for the UK CPI, which is given in the ONS CPI Technical Guide [1], and the international specification, which is set out in the International Labour Organisation CPI Technical Guide [2].

8.1 Specifying the index

We start with four important aspects of the index that need to be specified:

The geographical scope of the index. This is the whole of the UK, but not the Channel Islands, nor the Isle of Man, which are Crown dependencies.[1]

The reference population. This is all households and foreign visitors within the UK as well as residents of communal establishments such as university halls of residence.

---

[1] These are self-governing dependencies of the UK Crown.

A Practical Introduction to Index Numbers, First Edition. Jeff Ralph, Rob O'Neill and Joe Winton.
© 2015 John Wiley & Sons, Ltd. Published 2015 by John Wiley & Sons, Ltd.
Companion Website: http://www.wiley.com/go/ralph/index_numbers

The expenditure that should be included. This is the expenditure by the reference population.

The type of transactions that should be included. This is goods and services bought by the reference population for the purposes of consumption; this excludes expenditure on savings and investments.

These answers set some basic limits on what is included; the following sections look at a range of further aspects that need to be considered – we start with the goods and services for which we should be capturing prices.

8.2   The basket

The CPI aims to measure the change in the level of prices for consumer goods and services. One way to measure this would be to record the price for all transactions for all goods and services bought by consumers over two time periods: a price reference period, say, January 2014, and a comparison period, say, August 2014. This is, of course, an entirely impractical approach. Firstly, the number of distinct goods and services bought in any one month is enormous and secondly, the number of transactions is also huge. Clearly, a much simpler alternative is needed to make the measurement of price change possible within sensible time and cost limits. The approach taken is to take a sample of goods and services and a sample of transactions from a sample of locations.

The sample of goods and services is chosen to be representative of consumer purchases, so it covers a wide range of types of goods and services; for example, bananas, petrol, formal shirts, flower delivery and haircuts. In practice, the sample consists of about 700 items. This sample can be thought of as a basket of goods and services, which was introduced in Chapter 1 [3].

The items that make it into the basket are chosen according to a set of rules. Firstly, they must be items on which the consumers spend a significant amount of money. Secondly, an item may be chosen to represent a number of related items so as to keep the number of goods and services in the basket manageable. For example, it is not necessary to include both garden forks and spades; their price movements are assumed to be similar. If an area of goods and services is thought to be over-represented, an item will be removed.

Constructing a representative basket is not an activity that is performed once and is expected to remain fixed for many years. New goods and services appear and some disappear; in addition, consumer tastes and preferences change over time. These factors need to be reflected in the basket of goods. In the UK, the basket is updated every year in January [4]. Examples of recent changes are:

- Hair dryers out and hair straighteners in (2010)
- Lipstick out and lip gloss in (2010)
- Internet dating in (2013)
- Flavoured milk in (2014).

For items like bread, there are hundreds of types, sizes and brands available to consumers. Only a small selection of 'bread items' can be included, so the price movements of the large number of bread products are represented by a few carefully chosen types.

8.3   Locations and outlets

Besides taking a sample of goods and services from the vast number on offer in the marketplace, a sample of locations and shops, or outlets, must also be taken, as price collection cannot take place in every retail outlet in every town. The sampling is carried out in stages, with locations chosen first followed by outlets within the locations. About 140 locations and 20 000 shops (a location can contain more than one shop) are selected to represent the UK retail environment. Prices are also collected via phone, email and the Internet.

To reflect the changing nature of the retail environment, a subset of locations and outlets is changed every year; this is not only due to some retail outlets closing and new ones opening but also to rotate some of the sample, which is an important aspect of statistical best practice. The details of the procedure are contained in the ONS Consumer Prices Technical Manual [1].

Some retailers operate a national pricing policy, so prices are collected only once. Prices are also collected from the Internet because an increasing proportion of purchases are made online.

8.4   Price collection

Having chosen a sample of goods and services ('the basket') and a sample of locations and retail outlets, price collectors can then visit the selected outlets and capture prices for the selected items, with multiple prices collected for each specific item. In total, about 180 000 prices are recorded every month. We saw from Chapter 5 that the formula for the Lowe price index requires prices of goods and services to be collected at both the price reference period ('0') and the comparison period ('t'). For example the price reference period for 2014 is January and the comparison period is each subsequent month, as the index is produced monthly. As the task is to measure price change, it is important to take the price of comparable commodities at the two time periods. Price collectors, who visit outlets, are given descriptions of the items to price (e.g. 'white sliced loaf branded 800g') and will try to choose the product that is most representative of that item in the outlet. They record additional information about the chosen product and will always attempt to collect a price for the exact same item they priced in the previous month.

8.5   Weighting

The basket of goods and services contains a selection of representative items that cover the span of commodities that consumers buy. Once the basket has been specified, price change for each of the 700 items can be recorded. A simple approach to estimate the overall average price change would be to take an average of the individual price changes; however, it is readily apparent that consumers do not spend equal amounts on these items. A fairer measure of average price change is achieved by weighting the price changes of items in the basket by the proportion of overall expenditure that consumers spend on them (or on the groups of goods or services that the item represents). We saw in Chapter 5 that price indices such as the Laspeyres and Lowe can be written both in terms of 'prices and quantities' or alternatively in terms of 'price relatives and expenditure shares (weights)'. It is the latter representation that is used in practice in compiling the CPI.

The proportions of expenditure on different types of goods and services is specified each year in the UK by a set of weights published at the start of the calendar year by the Office for National Statistics [5]. The weights are mainly obtained from the Household Final Monetary Consumption Expenditure component of the National Accounts [1] and a living costs and food survey; weights are specified in parts per thousand. For example, carpets and other floor coverings had a weight of four parts per thousand in 2014 while meat had a weight of 21, indicating that the expenditure on meat was more than five times the expenditure on carpets and other floor coverings. An example of the weights was given in Chapter 1 for 2013. The changes in weights from year to year reflect the changing consumer spending patterns, which is interesting in itself; for example, the weight for 'housing, water, electricity and gas' for 2013 was '244' and for 2014, it was '265' showing an increased expenditure on this class of goods and services.

8.6   Aggregation structure

We saw in Chapter 6 that an index can be built up from smaller, domain indices. This has the advantage of producing lower-level indices, which provide the building blocks for higher level indices. In this way, an overall index is constructed in stages in a tree-like structure. This is the approach followed for the CPI.

The tree-like structure used in the CPI is a classification system – an international classification system of household consumption expenditure known as 'COICOP' – the Classification of Individual Consumption by Purpose. This was introduced in Chapter 6. The single, headline inflation figure, the all-items level, is at the top of the tree. The all-items figure is divided into 12 divisions with numerical codes for each domain; for example, 'Food and Non-Alcoholic Beverages' (01) and 'Clothing and Footwear' (03). Index numbers for these 12 divisions are published as they are of interest to users. These index numbers of the 12 divisions are combined together with expenditure weights to form the all-items index number. Underneath these divisions are groups, for example 'Food' (01.1) and 'Non-alcoholic beverages' (01.2). Below this are classes, for example 'Bread and cereals' (01.1.1); this is the lowest level at which indices are published. Below classes are items.

Underneath the item level, we reach the lowest weighted level of the structure, known as the stratum level. At this level, we reach very specific types of goods and services. For example, a stratum is 'white sliced 800 g loaves bought from independent stores in the South East region of England'. An item index is calculated from stratum indices and stratum weights. Each stratum index is known as an elementary aggregate within which price relatives are combined without weighting information, as expenditure weights are not available at such a low level.

The ability to calculate index numbers in stages and combine them up the classification structure is the result of adopting a linear formula, the Lowe index. As was noted in Section 6.7, if a non-linear index formula had been chosen to combine price relatives and weights, then a different process would be needed to ascend the aggregation structure. To calculate an index at the class level, all the constituent price relatives and weights from the item level to the elementary aggregates that make up that class would need to be combined.

8.7   Elementary aggregates

The very lowest level of the aggregation structure comprises price indices for single, relatively homogeneous goods such as the white sliced 800 g loaf described in Section 8.6. For some items where purchasing patterns differ according to the type of shop or the region, the price samples are stratified further to improve the quality of the indices [6]. The four stratum types are:

- Region and shop type

- Region only

- Shop type only

- No stratification.

In calculating the CPI, two shop types are used: multiples and independents. Independents are retailers with fewer than 10 outlets; regions are geographic areas such as the 'North East of England'. These strata are then combined to produce an index for each item, with stratum weights assigned according to the relative weights of expenditure across regions and shop types. In theory, all items for which prices are collected locally should be stratified by both region and shop type; however, this requires effective weighting information and is not always available [1]; thus the approach taken is to stratify as far as weighting information is reliable.

An example of a doubly stratified item is the previous example of 'white, sliced, 800 g loaves of bread bought from independent stores in the South East region of England'. There are about 5000 elementary aggregates in total.

At this level, weights for each price relative are not available[2] and price indices for each elementary aggregate are calculated using prices alone; that is, prices are

---

[2] However, in some cases where more information is available, weights are used at the price quote level. An example is air fares where some information on expenditure is known.

combined using un-weighted price index formulae as described in Chapter 4. There are two formulae used to combine price information in the UK CPI:

- The geometric mean or Jevons index[3]
- The ratio of arithmetic averages of price relatives or the Dutot index.

The choice of elementary aggregate formula is a complex and controversial matter, both from a technical point of view and from a historical one. It has led to much debate over the differences between the UK CPI and the UK Retail Prices Index (RPI), which uses a combination of Dutot and Carli indices.

Mathematically, the geometric mean of price relatives is always less than or equal to the arithmetic mean, so the RPI (which makes greater use of the arithmetic mean) tends to take a higher value than the CPI (which uses the geometric mean) at the same time period.[4] The value of the difference resulting from using a different formula at the elementary aggregate level has become known as 'the formula effect'. The size of the formula effect has varied over time and is affected by the distribution of price relatives in an elementary aggregate (known as 'price relative dispersion'), with a larger formula effect as the spread of price relatives increases.

The choice of elementary aggregate formula was the subject of a consultation by the UK National Statistician in 2012–2013, where consideration was given to removing the Carli formula from the RPI and replacing it with the Jevons formula; this would have had the effect of lowering the RPI. However, it was decided to not change the RPI formula, but to create a new index, the Retail Prices Index with a Jevons formula (RPIJ) [7].

There has been much study to try to identify the best formula to use where weights are not available [8, 9]; however, recent research suggests that it is not possible to emulate a weighted formula with an un-weighted one in an accurate manner [10]. It is an area of the construction of price indices that might be improved by the use of alternative sources of price and expenditure data in future. In the meantime, the best approach that can be taken is to follow the current international guidance [11].

The Carli formula is effectively forbidden in the Harmonised Index of Consumer Price (HICP) regulations as it tends to produce index numbers that are significantly different from the Dutot and Jevons formulae [12]. However, not all commentators agree with the justification for this ruling.

8.8   Linking

Chapter 7 introduced the chain-linking technique; we saw that this is a method for preserving the continuity of a time series of index numbers while allowing for aspects of the data and the methods to change. This technique is used in the CPI to accommodate changes in the consumer economy, thereby keeping the index 'relevant' to

---

[3] About 70% of the CPI is created using geometric means.
[4] There are other differences in methodology, which means that the RPI can be lower than the CPI; for example, the treatment of costs associated with housing.

current conditions. As has been noted before, in the case of the CPI, we update the basket and weights every year.

It used to be the case that National Statistics Institutes (NSIs) would make these changes every 5–10 years. However, this relatively long interval was considered to miss important changes, such as the introduction of new technology, and gradually, the NSIs have switched to making changes every year. In the UK, the index is unusual in that it is chain-linked twice – once in January of each year to take account of the annual update in weights and once more in February to account for the update of the basket.

The use of a chain-linked index does have implications for comparisons across the chain link. For comparisons between two months in the same year, say, September and March, the basket and weights are the same and it is just the prices that change. However, if a comparison is made between the September of year y and the September of year $y - 1$, then it is not just prices that have changed, but the basket and weights too. This is not a pure comparison of price changes as one might expect for a price index; instead, it is the price change for an adjusted, representative basket and the accompanying weights. Comparison across multiple years means comparison across many changes in the basket and weights.

8.9   Owner occupier housing

The treatment of costs related to owning a house is complex and the CPI does not include these expenditures, which comprise the costs of buying or owning a property, or the cost of maintenance such as replacing the roof or installing a new kitchen. This has been recognised as a significant omission and work has been underway for many years to develop an appropriate measure. Of course, a house (or flat) is not consumed in the way a bar of chocolate is consumed. Instead, it is considered to provide a flow of services to the owner over a long period of time; it is these services that are 'consumed' by an owner occupier. There are a number of theoretical approaches to accounting for these costs and the UK has decided to adopt the rental equivalence method, which works out how much it would cost to rent an equivalent house. This method says that the rental cost of a property is equal to the cost of the flow of services that an owner-occupied property provides to its owner.

To accommodate the development of a measure for owner-occupied housing, a new index was published in 2013, the CPIH, which is similar in its construction to the CPI but also includes owner-occupied housing costs. By including these costs, the expenditure shares (i.e. the weights) for all commodities become different; this is accounted for in the production of the CPIH [13].

8.10   Publication

The CPI together with the RPI, CPIH and their sub-families of indices is published monthly around the middle of a calendar month and relates to prices collected in the previous month; the publication dates are announced six months in advance. The monthly statistical release consists of a short summary, a statistical bulletin and a set of data tables.[5] The statistical bulletin provides a commentary on the figures and identifies the significant changes over the past month and the past year. Standardised data tables can be downloaded by users for further analysis.

8.11   Special procedures

Housing is not the only item that needs to have particular treatment applied to reflect its complex nature. Special consideration is also needed for a range of goods and services, including university fees, the London congestion charge, airfares, mobile phone costs and electricity and gas tariffs. The details are covered in the ONS Technical Manual [1].

8.12   Chapter summary

This chapter has given a very brief overview of the how the CPI is constructed; it is a complex process that has to contend with many challenges. It should also be noted that at several points, there is a trade-off to be made in the construction of the CPI between what might be best to do from a theoretical standpoint and the practical considerations of calculating index numbers to tight deadlines. When we consider the theory of how to construct a price index, we are freed from such constraints; however, we should keep in mind that these practical considerations are vital for National Statistical Institutes who produce the index numbers.

References

1. See ONS (2014) *Consumer Price Indices Technical Manual*, Office for National Statistics, http://www.ons.gov.uk/ons/guide-method/user-guidance/prices/cpi-and-rpi/cpi-technical -manual/consumer-price-indices-technical-manual--2014.pdf (accessed 16 January 2015).
2. See ILO (2004) *Consumer Price Index Manual: Theory and Practice*, International Labour Organisation, Geneva. http://www.ilo.org/public/english/bureau/stat/guides/cpi/ #manual (accessed 16 January 2015).
3. Office for National Statistics (2013) *Consumer Price Indices A Brief Guide*, Office for National Statistics, http://www.ons.gov.uk/ons/guide-method/user-guidance/prices/cpi- and-rpi/consumer-price-indices--a-brief-guide.pdf (accessed 16 January 2015).
4. Office for National Statistics (2014) *Consumer Price Inflation Basket of Goods and Services, 2014*, Office for National Statistics, http://www.ons.gov.uk/ons/guide-method/user -guidance/prices/cpi-and-rpi/cpi-and-rpi-basket-of-goods-and-services/consumer-price -inflation-basket-of-goods-and-services-2014.pdf (accessed 16 January 2015).

---

[5] See, for example, the August 2014 release at http://www.ons.gov.uk/ons/rel/cpi/consumer-price -indices/august-2014/index.html.

5. Jenkins, C. and Bailey, S. (2014) Office for National Statistics, Consumer Price Inflation Statistics – Updating Weights for 2014, http://www.ons.gov.uk/ons/rel/cpi/cpi-and-rpi-index--updating-weights/2014/index.html (accessed 16 January 2015).

6. See Office for National Statistics (2014) *Consumer Price Indices Technical Manual*, Office for National Statistics, p. 42, http://www.ons.gov.uk/ons/guide-method/user-guidance/prices/cpi-and-rpi/cpi-technical-manual/consumer-price-indices-technical-manual--2014.pdf (accessed 16 January 2015).

7. Office for National Statistics (2013) *National Statistician Announces Outcome of Consultation on the RPI*, Office for National Statistics, http://www.ons.gov.uk/ons/rel/mro/news-release/rpirecommendations/rpinewsrelease.html (accessed 16 January 2015).

8. See Chapter 20 in ILO (2004) *Consumer Price Index Manual: Theory and Practice, Elementary Indices*, International Labour Office, http://www.ilo.org/public/english/bureau/stat/guides/cpi/#manual (accessed 16 January 2015).

9. Levell, P. (2014) Is the Carli index flawed?: assessing the case for the new retail price index RPIJ. *Journal of the Royal Statistical Society, Series A*, **178** (2), 1–34.

10. Elliott, D., Winton, J., and O'Neill, R. (2013) Elementary aggregate indices and lower level substitution bias. *Statistical Journal of the IAOS*, **29** (1), 11–19.

11. See Chapter 18:ILO (2014) *The Consumer Price Index Manual: Theory and Practice*, International Labour Office, http://www.ilo.org/public/english/bureau/stat/guides/cpi/#manual (accessed 16 January 2015).

12. Eurostat (2014) HICP Methodology, Elementary Aggregates, http://epp.eurostat.ec.europa.eu/statistics_explained/index.php/HICP_methodology#Elementary_aggregate_indices (accessed 16 January 2015).

13. Office for National Statistics (2014) *Consumer Price Inflation Detailed Briefing Note*, Office for National Statistics, http://www.ons.gov.uk/ons/rel/cpi/consumer-price-indices/september-2014/article-briefing-note.html (accessed 16 January 2015).

Exercise G

These questions are designed to encourage you to think more broadly about the topics you have covered and how they can be applied in practice. Formal solutions are not provided; however, some supporting information can be found in the online content that accompanies this book.

G.1 Imagine you are a price collector who has been given a list of food items to price; what difficulties might you face?

G.2 Would the difficulties of collecting prices be increased, if you were then asked to collect information on women's dresses?

G.3 What items might have been in the CPI basket of goods 10 years ago but are now not widely used? What about 50 years ago?

G.4 If you were compiling a price index for the basket of goods and services that you buy, what items would be included?

G.5 What are the potential problems with having to compile the CPI for the entire UK?

G.6 How many shop types or shopping locations can you think of in your own area? Which are the most important to be sampled for a price index and why?

9

Re-referencing a series

One of the key uses of index numbers is in making comparisons of changes in data series that are not otherwise easily compared. For example, how does the change in the price of a pint of beer compare to the change in the price of a television? In this chapter, we demonstrate how re-referencing a series can improve the quality of such comparisons.

9.1 Effective comparisons with index numbers

Table 9.1 shows the price of a pint of beer and a given television set over the same 12-month period; these figures are then plotted in Figure 9.1.

From the table, we can see the prices of both the television and the pint of beer change over the year. Graphs are often an effective way of comparing movements over time but according to Figure 9.1, in which we plot the levels of the two price series, it is very hard to see any movement in the price of beer at all. In addition, the difference in magnitude of the price of the two items makes a comparison of the two items difficult. In Table 9.2, these prices have been converted to index series with January set as the base period taking the standard value of 100.

If we now plot the price indices, as in Figure 9.2, we see a much clearer picture. We see now that the price of the television has a general downward trend over the year with some fluctuation, whereas the price of the beer rises by a third over the course of the year.

The above example illustrates how useful index numbers can be in comparing data series. Now look at Figure 9.3. This chart shows the Gross Value Added (GVA) for different sectors of the UK economy over an 11-year period. GVA is a measure of the value of goods and services produced in an area, industry or sector of an economy and is a key part of Gross Domestic Product.

A Practical Introduction to Index Numbers, First Edition. Jeff Ralph, Rob O'Neill and Joe Winton.
© 2015 John Wiley & Sons, Ltd. Published 2015 by John Wiley & Sons, Ltd.
Companion Website: http://www.wiley.com/go/ralph/index_numbers

Table 9.1   Beer and television prices, January–December.

			Price (£)			
	January	February	March	April	May	June
Television	350.00	350.00	340.00	342.00	335.00	350.00
Beer	2.25	2.30	2.30	2.30	2.35	2.35
	July	August	September	October	November	December
Television	365.00	350.00	325.00	325.00	310.00	310.00
Beer	2.40	2.50	2.80	2.80	2.80	3.00

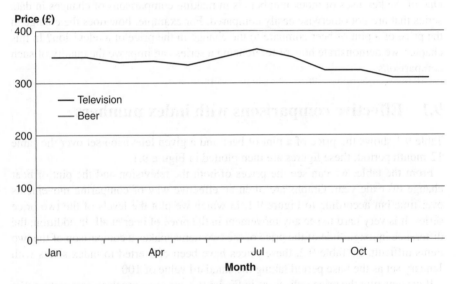

Figure 9.1   Beer and television prices, January–December.

Table 9.2   Beer and television price indices, January–December (January = 100).

			Index (January = 100)			
	January	February	March	April	May	June
Television	100.0	100.0	97.1	97.7	95.7	100.0
Beer	100.0	102.2	102.2	102.2	104.4	104.4
	July	August	September	October	November	December
Television	104.3	100.0	92.9	92.9	88.6	88.6
Beer	106.7	111.1	124.4	124.4	124.4	133.3

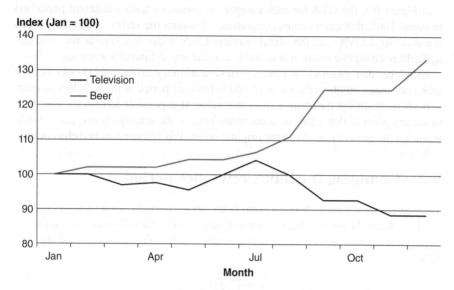

Figure 9.2   Beer and television price indices, January–December.

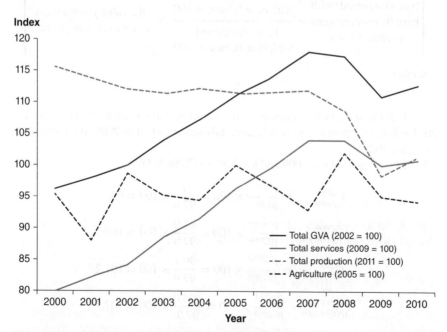

Figure 9.3   Gross value added (GVA) chained volume measure at basic prices United Kingdom, by category of output.

In Figure 9.3, the GVA for each category is presented with a different period set to equal 100[1]; this complicates comparisons between the series. Although we can see that 'Total GVA' and the 'Total Services GVA' move in a similar way and that agriculture changes as well, it is difficult to make any definitive statements.

The values that we give to our index series are arbitrary; so far in this book, we have given our index numbers the value of 100 in the base period with the values in other periods representing the change from this value. It is possible to re-scale the series to set any point in that series to equal some value – we usually choose 100 – while retaining the growth rates from the original series; this is known as **re-referencing**.

9.2   Changing the index reference period

To re-reference a series, choose a point in the series, *period r,* for which you wish to set the index to be some value α – we will call *period r* the reference period; then at every point in the series, divide the index number by the value at *period r* and multiply by α.

$$I^{0,t}_{(r=\alpha)} = \frac{I^{0,t}}{I^{0,r}} \times \alpha \tag{9.1}$$

$$\left(\begin{array}{c} \text{Index for period } t \text{ with} \\ \text{base 0, re-referenced} \\ \text{to period } r = \alpha \end{array} = \frac{\begin{array}{c} \text{Index for period } t \\ \text{with base 0, base} = 100 \end{array}}{\begin{array}{c} \text{Index for period } r \\ \text{with base 0, base} = 100 \end{array}} \times \begin{array}{c} \text{the value you want the} \\ \text{series to take at period } r \end{array} \right)$$

So that:

$$I^{0,r} = \alpha$$

Look at the short index series in Table 9.3. This index is currently referenced to $2011 = 100$. It is possible to re-reference this series so that in 2009, the index takes the value 100.

In this case, period r is 2009 and $\alpha = 100$, so (Table 9.4):

$$I^{0,2008}_{(2009=100)} = \frac{I^{0,2008}}{I^{0,2009}} \times 100 = \frac{96.7}{97.9} \times 100 = 98.7$$

$$I^{0,2009}_{(2009=100)} = \frac{I^{0,2009}}{I^{0,2009}} \times 100 = \frac{97.9}{97.9} \times 100 = 100.0$$

$$I^{0,2010}_{(2009=100)} = \frac{I^{0,2010}}{I^{0,2009}} \times 100 = \frac{99.1}{97.9} \times 100 = 101.2$$

$$I^{0,2011}_{(2009=100)} = \frac{I^{0,2011}}{I^{0,2009}} \times 100 = \frac{100.0}{97.9} \times 100 = 102.1$$

$$I^{0,2012}_{(2009=100)} = \frac{I^{0,2012}}{I^{0,2009}} \times 100 = \frac{100.8}{97.9} \times 100 = 102.9$$

---

[1] The data have been presented in this way by the authors to illustrate the benefits of re-referencing a series.

Table 9.3   Short index series (2011 = 100).

Year (t)	2008	2009	2010	2011	2012
Index $\left(I^{0,t}_{(2011=100)} \right)$	96.7	97.9	99.1	100.0	100.8

Table 9.4   Short index series (2011 = 100 and 2009 = 100).

Year (t)	2008	2009	2010	2011	2012
Index $\left(I^{0,t}_{(2011=100)} \right)$	96.7	97.9	99.1	100.0	100.8
Re-referenced index $\left(I^{0,t}_{(2009=100)} \right)$	98.7	100.0	101.2	102.1	102.9

In practical terms, the value (or level) of the index series does not matter, and what we are most concerned with is the growth rate. Multiplying all of the values in the series by a constant will not have any impact on any growth rate.

Recall from Chapter 2 that the growth rate in an index series between period s and period t is defined as:

$$g^{s,t} = \frac{(I^{0,t} - I^{0,s})}{I^{0,s}} \times 100 \tag{9.2}$$

We can show that the growth rate in our original series is the same as the growth rate in our re-referenced series. We can express the re-referenced series as:

$$I^{0,t}_{(r=\alpha)} = I^{0,t} \times \beta \tag{9.3}$$

where

$$\beta = \frac{\alpha}{I^{0,r}} \tag{9.4}$$

so that the growth rate in the re-referenced series is:

$$\begin{aligned}
g^{s,t}_{\text{re-referenced}} &= \frac{(\beta \cdot I^{0,t} - \beta \cdot I^{0,s})}{\beta \cdot I^{0,s}} \times 100 = \frac{\beta \cdot (I^{0,t} - I^{0,s})}{\beta \cdot I^{0,s}} \times 100 \\
&= \frac{(I^{0,t} - I^{0,s})}{I^{0,s}} \times 100 = g^{s,t}_{\text{Original}}
\end{aligned}$$

We can further show this using the index series presented in Table 9.4. The growth rate in the original series (2011 = 100) between 2008 and 2012 is:

$$g^{s,t} = \frac{(I^{0,t} - I^{0,s})}{I^{0,s}} \times 100 = \frac{(100.8 - 96.7)}{96.7} \times 100 = 4.255$$

The growth in the re-referenced series (2009 = 100) between 2008 and 2012 is:

$$g^{s,t}_{\text{re-referenced}} = \frac{(\beta \cdot I^{0,t} - \beta \cdot I^{0,s})}{\beta \cdot I^{0,s}} \times 100 = \frac{(102.9 - 98.7)}{98.7} \times 100 = 4.255$$

9.3   Why re-reference?

We have just seen that it is possible to rescale an index series without affecting the movements in the series. There are two main reasons due to which you might re-reference a series. The first is that the further away from the reference period that the series gets, the more difficult it is to interpret movements in the series. Take a look at Table 9.5; in the original series, we can see movements up and down, but we would have to do a bit of work to get more information. By re-referencing the series to set the February value equal to 100, we can see a much clearer picture; for example, we can see that the series has grown by 3.9% between July and February. Re-referencing an index series in this way can make it easier to see the underlying pattern in a series.

The second reason for re-referencing is to enable comparisons between different series. Recall the GVA example at the start of this chapter; we wanted to compare four series showing GVA for different sectors of the UK economy. It was difficult to make comparisons between the series as they were all referenced to different periods. In Figure 9.4, the four measures of GVA have been **re-referenced** so that 2003 = 100 for all of the series.

Table 9.5   A simple series.

	February	March	April	May	June	July	August
Original	425.5	426.2	432.0	428.9	430.2	442.3	443.5
Re-referenced	100.0	100.2	101.5	100.8	101.1	103.9	104.2

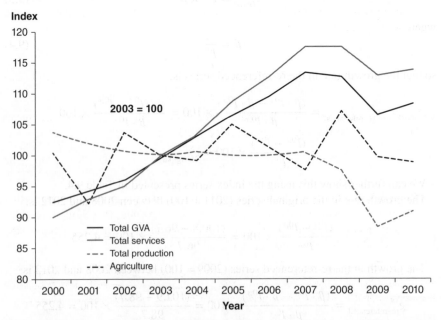

Figure 9.4   Gross value added (GVA) chained volume measure at basic prices United Kingdom, by category of output.

Re-referencing the four series to a common time period allows us to make a clearer comparison of the growth in the four series. We can now see that while 'Total GVA' and 'Total Services' move in a similar way, the GVA for the services sector grows at a faster rate over the period, whereas the GVA for the production sector falls over the same period.

It is important to note that re-referencing does not change the base period. Each value in a re-referenced index series still represents the change in the data series between the base period and the current period, but the value that the index series takes at the base period is no longer equal to 100.

9.4   Re-basing

There are three reference periods that are used in index numbers:

- The index reference period

- The price reference period

- The weight reference period.

The *index reference period* is the time period where the index is set to equal 100; the process of re-referencing changes this time period.

The *price reference period* is the time period with which we make comparison – it is the time period '0' in p_0.

The *weight reference period* is the time period from which the weights are drawn.

These time periods can be different. For example, for the Consumer Prices Index, the current index reference period is 2005, the current price reference period is January 2014 and the weight reference period is the calendar year 2012, though the weights (at the class level) are 'price updated' to December 2013 [1]. Note that although the Consumer Prices Index is a monthly index, the index reference period and the weight reference periods are a year.

Sometimes, more than one of these time periods is changed at the same time. *Re-basing* is a generic term that is used to describe the updating of one or more of these time periods. It is often used to describe the process by which all three reference periods are updated. For example, the UK Services Producer Price Index (SPPI) was 're-based' in February 2014. The index reference period was changed from 2005 = 100 to 2010 = 100, the weight reference period was changed from 2005 to 2010 and the price reference period from the first quarter of 2005 to the first quarter of 2010. For some service industries, the weight reference period was updated to a time between 2005 and 2010 [2].

Because the term 're-basing' is a generic description, the exact meaning has to be taken from the context. It is also common for other changes, such as changes to classifications and the methods to be included at the same time. For example, in the case of the SPPI re-basing described above, changes to the aggregation structure were also made.

9.5   Chapter summary

In this chapter, we have considered how we can use a straightforward mathematical technique to re-reference an index series and, therefore, make it easier to compare data series. This is an illustration of one of the key advantages of using index number representations of data.

We have also considered the difference between re-referencing and re-basing of an index. The two operations are important when using and compiling indices; however, it is important to be clear about the differences between them and to be aware that they are employed for distinct purposes.

References

1. Office for National Statistics (2014) Consumer Prices Index and Retail Prices Index: Updating Weights, 2014, http://www.ons.gov.uk/ons/rel/cpi/cpi-and-rpi-index--updating-weights/2014/index.html (accessed 14 January 2015).
2. Office for National Statistics (2014) The Impact of Rebasing the Services Producer Price Index to 2010 = 100, http://www.ons.gov.uk/ons/rel/ppi2/services-producer-price-indices/services-producer-price-indices-rebasing--2010-100-/impact-of-rebasing-the-services-producer-price-index-to-2010-100.html (accessed 14 January 2015).

Exercise H

Solutions to this exercise can be found in Appendix D.

H.1 Using the data in Table H.1, construct a Carli index series of transport costs with 2007 as the base year and the index value in 2007 equal to 100.

Table H.1   Transport costs 2007–2014.

	Bus	Train	Taxi
2007	1.2	10	23
2008	1.8	11	25
2009	1.9	14	28
2010	2.2	16	29
2011	3.3	17	32
2012	4.5	18	34
2013	4.6	21	38
2014	4.8	22	40

H.2 With the index series you calculated above, re-reference the series so that 2012 has an index value of 100.

H.3 Show that the percentage change in the index number you calculated in question H.1 is the same as that in question H.2 when comparing costs in 2009 with costs in 2013.

H.4 Recalculate the Carli index of transport costs with 2010 as the base period and then re-reference the index to 100 in 2010.

H.5 Show that the percentage change between 2009 and 2013 from the index calculated in question H.4 is different to either your original or re-referenced indices from questions H.1 and H.2.

10

Deflation

The relationship between price and quantity is central to index number theory and practice in the economic sphere. For any item, the value can be calculated as the unit price of that item multiplied by the quantity. As we have seen in earlier chapters, the problem of decomposing the change in the overall value of a basket of goods into some combination of a price index and a quantity index for the same period (in a 'fair' way) is known as the Index Number Problem, which has been a consistent issue in the discussion of index numbers over many years.

We saw in Section 5.5 that the Laspeyres and Paasche indices provide solutions. The change in value of a basket of goods between two periods can be expressed as the product of a Laspeyres price index and a Paasche quantity index or as the product of a Paasche price index and a Laspeyres quantity index between the same periods. The same is also true of the Fisher index.[1]

$$V^{0,t} = P^{0,t}_{\text{Laspeyres}} \cdot Q^{0,t}_{\text{Paasche}} = P^{0,t}_{\text{Paasche}} \cdot Q^{0,t}_{\text{Laspeyres}} \tag{10.1}$$

We have also seen in earlier chapters how a price index can be constructed from collected prices and expenditure shares. Calculating a quantity index is not so straightforward. In practice, although both prices and values are relatively easy to capture, quantities are not. Consider capturing this information from shops. Shops will display their prices and will be able to tell you their total revenue; however, asking them to report on how many of each item they have sold would require a significant amount of effort for them. The relationship between Paasche and Laspeyres indices can be used to obtain a quantity index by using the appropriate value and price indices in the following way:

$$Q^{0,t}_{\text{Paasche}} = \frac{V^{0,t}}{P^{0,t}_{\text{Laspeyres}}} \quad \text{or} \quad Q^{0,t}_{\text{Laspeyres}} = \frac{V^{0,t}}{P^{0,t}_{\text{Paasche}}} \tag{10.2}$$

[1] You might try proving these statements to yourself; proofs are provided in Appendix A.

A Practical Introduction to Index Numbers, First Edition. Jeff Ralph, Rob O'Neill and Joe Winton.
© 2015 John Wiley & Sons, Ltd. Published 2015 by John Wiley & Sons, Ltd.
Companion Website: http://www.wiley.com/go/ralph/index_numbers

Note that if price and value indices are expressed with the base period equal to 100 rather than 1, then to express the quantity index in the same way, we need to multiply by 100:

$$Q^{0,t}_{\text{Paasche}} = 100 \times \frac{V^{0,t}}{P^{0,t}_{\text{Laspeyres}}} \quad \text{or} \quad Q^{0,t}_{\text{Laspeyres}} = 100 \times \frac{V^{0,t}}{P^{0,t}_{\text{Paasche}}} \qquad (10.3)$$

We have seen in earlier chapters why a price index is important to measure; Chapter 3 described a wide variety of uses for it. One way of describing a price index is as a means of converting the value of money from one period of time to another, which is a very important activity. But what about quantity indices – why are they important? A quantity index for a collection of goods and services is a measure of the overall volume of items, which combines quite different goods and services. The Retail Sales Index provides a useful example here. We considered the total value of retail sales at the start of Chapter 1. We know that the overall change in the value of retail sales is made up of two separate types of change – a change in the overall price of retail goods and a change in the overall volume (or quantity) of retail goods sold. It is interesting to see the change in the value of retail sales over time; however, this includes the change in the value of money as well as the quantity of goods sold. The time series of the retail sales quantity index tells us something important as well – how the volume of sales is changing. The retail sales statistical bulletin from the Office for National Statistics provides the values of retail sales over time and the volume of retail sales as well.

Example 10.1   Table 10.1 shows price and value information for a bakery in 2012. For each month in 2012, we have the change from January in the value of cakes sold each month as well as the change from January in the price of cakes sold each month as measured by a Laspeyres price index.

Table 10.1   Change in price and value of cakes sold by a bakery in 2012.

Period	Change in value $V^{\text{Jan 2012},t}$	Laspeyres price index $P^{\text{Jan 2012},t}_{\text{Laspeyres}}$
January	100.0	100.0
February	101.0	100.5
March	101.6	99.5
April	91.0	87.6
May	104.5	100.5
June	100.8	96.9
July	98.0	91.5
August	97.6	87.1
September	110.1	101.6
October	110.2	98.1
November	111.6	101.7
December	112.0	101.9

The owner of the bakery would like to measure the change in the overall volume of cakes sold over 2012. Using the information in Table 10.1, calculate an appropriate quantity index for each period with January 2012 as the base period.

Solution We have a value index and a Laspeyres price index; we can use these to calculate a Paasche quantity index as follows:

For February 2012:

$$Q_{Paasche}^{Jan,Feb} = 100 \times \frac{V^{Jan,Feb}}{P_{Laspeyres}^{Jan,Feb}} = 100 \times \frac{101.0}{100.5} = 100.5$$

And similarly for the rest of the series:

Period	Change in value $V^{Jan\ 2012,t}$	Laspeyres price index $P_{Laspeyres}^{Jan\ 2012,t}$	Paasche quantity index $Q_{Paasche}^{Jan\ 2012,t}$
January	100.0	100.0	$100/100 \times 100 = 100.0$
February	101.0	100.5	$101/100.5 \times 100 = 100.5$
March	101.6	99.5	$101.6/99.5 \times 100 = 102.1$
April	91.0	87.6	$91/87.6 \times 100 = 103.9$
May	104.5	100.5	$104.5/100.5 \times 100 = 104.0$
June	100.8	96.9	$100.8/96.9 \times 100 = 104.0$
July	98.0	91.5	$98/91.5 \times 100 = 107.1$
August	97.6	87.1	$97.6/87.1 \times 100 = 112.1$
September	110.1	101.6	$110.1/101.6 \times 100 = 108.4$
October	110.2	98.1	$110.2/98.1 \times 100 = 112.3$
November	111.6	101.7	$111.6/101.7 \times 100 = 109.7$
December	112.0	101.9	$112/101.9 \times 100 = 109.9$

10.1   Value at constant price

Deflation is the process of removing the effect of price changes from value data. Using the relationship between Laspeyres and Paasche price and volume indices, it is possible to derive a measure of value at **constant price.** This process is carried out for a number of economic statistics; for example, when GDP (Gross Domestic Product) is reported in year 2002 pounds, this means that the GDP value series has been deflated to account for price change.

Current period values can be deflated by dividing by a Paasche price index or multiplying the base period value by a Laspeyres quantity index to produce a measure of value where prices are held constant at the base period but quantities relate to the current period; such a measure is known as a **volume measure**[2]:

$$v^t \big/ P^{0,t}_{\text{Paasche}} = v^0 \cdot Q^{0,t}_{\text{Laspeyres}} = \sum p^0 q^t \qquad (10.4)$$

that is:

$$v^t \big/ P^{0,t}_{\text{Paasche}} = \frac{\sum p^t q^t}{\sum p^t q^t \big/ \sum p^0 q^t} = \frac{1}{1 \big/ \sum p^0 q^t} = \sum p^0 q^t \qquad (10.5)$$

and:

$$v^0 \cdot Q^{0,t}_{\text{Laspeyres}} = \sum p^0 q^0 \times \frac{\sum p^0 q^t}{\sum p^0 q^0} = \sum p^0 q^t \qquad (10.6)$$

Value data are often known as current price or 'CP' data, and measures of value at constant price are known as constant price or 'KP' data.

A deflated value ratio can also be obtained by dividing by a Laspeyres price index to yield a Paasche quantity index. Given the difficulty in calculating Paasche price indices, it is often the case that deflation is carried out with Laspeyres price indices even though a Laspeyres volume index is the desired output. This is not strictly correct; however, in practice, deflation is often carried out at a detailed level where the difference between a Laspeyres and a Paasche index is expected to be small.

Note: as with deriving volume indices above, the effect of expressing indices as a percentage will need to be accounted for; so the formulae become:

$$v^t \big/ P^{0,t}_{\text{Paasche}} \times 100 = \sum p^0 q^t \qquad (10.7)$$

$$\frac{v^0 \cdot Q^{0,t}_{\text{Laspeyres}}}{100} = \sum p^0 q^t \qquad (10.8)$$

10.2   Volume measures in the national accounts

Statisticians and economists are often interested in two types of series:

- Current Year Prices (CYP)
- Constant Prices (KP)

A measure at **Current Year Prices** is a measure of value; prices and quantities both come from the current period.

A measure at **Constant Prices** is a volume measure where quantities come from the current period but prices come from some reference period, r; we might refer to this as a measure at period r prices.

---

[2] It is worth noting that while volume measures are usually expressed in pounds or dollars, they are not actually money amounts.

To do this, we either adjust the period t value for change in prices:

$$KP^{t;r} = v^t \frac{P^{0,r}_{Paasche}}{P^{0,t}_{Paasche}} = \frac{v^t}{\left(\dfrac{P^{0,t}_{Paasche}}{P^{0,r}_{Paasche}}\right)} \tag{10.9}$$

Or adjust the period r value for change in quantities:

$$KP^{t;r} = v^t \left(\frac{v^r}{v^0}\frac{1}{Q^{0,r}_{Laspeyres}}\right)\left(\frac{v^0}{v^t}Q^{0,t}_{Laspeyres}\right) = v^r \frac{Q^{0,t}_{Laspeyres}}{Q^{0,r}_{Laspeyres}} \tag{10.10}$$

10.3   Chapter summary

In this chapter, we have seen how the relationship between Price, Quantity and Value can be used to produce volume indices that would otherwise be too difficult to determine. We have also seen how to deflate data series using different indices. This technique is extremely useful, especially when we are faced with knowing only a value series and a price index series as we can then use them to remove the impact of price changes and provide an insight into the underlying economic activity. An interesting example of this occurs where revenues and prices are both increasing; however, the actual quantities sold are falling, which would suggest that demand is falling less than prices are increasing, implying that demand for a good may well be what economists refer to as inelastic.

Exercise I

Solutions to this exercise can be found in Appendix D.

I.1 A company has recorded the value of its sales over a number of years, which show an increase for all years; these are shown in the table below. Also shown are values of a Paasche price index relevant to the goods the company produces (Table I.1).

a. Calculate the value of sales data series at 2007 prices.

b. Calculate a Laspeyres quantity index series.

c. Has the increase in sales been the result of an increased volume of sales, or from the change in the prices?

d. Calculate a further series at 2010 prices.

Table I.1   Value of sales 2007–2012.

Year	Value of sales	Paasche price index
2012	23.54	1.432
2011	22.28	1.302
2010	21.99	1.271
2009	20.31	1.182
2008	19.76	1.101
2007	18.52	1.000

I.2 A data series has been provided in CYP together with a Laspeyres quantity index (Table I.2):

Calculate the value of sales at 2008 prices.

Table I.2   CYP and Laspeyres quantity index series, 2004–2009.

Year	Value of sales (CYP)	Laspeyres quantity index
2009	1487	0.978
2008	1465	0.974
2007	1452	0.968
2006	1398	0.943
2005	1382	0.921
2004	1366	0.906

11

Price and quantity index numbers in practice

Previous chapters have introduced the concept of index numbers, the theory behind price indices and, in particular, their application in quantifying inflation. This chapter provides an overview of the range of price and quantity indices produced to meet user needs. Although this chapter will describe UK statistical outputs, similar statistics are produced by most National Statistics Institutes (NSIs) around the world.

11.1   A big picture view of price indices

Price indices are some of the most widely used of all Official Statistics. They are produced to satisfy the needs of a wide range of users including central banks, Governments, businesses, researchers and citizens. For the general public, the most familiar price indices are consumer price indices; these measure the changes in the price of goods and services that households typically buy. In the UK, there are two main measures, the Consumer Prices Index (CPI) and the Retail Prices Index (RPI); as we saw in Chapter 3, they are used for a wide range of purposes, including as a guide for monetary policy, the indexation of wages, contracts, tax allowances, state benefits and pensions. They are also part of the price information used to remove the effects of inflation from important outputs such as the National Accounts.

To meet the wider user needs for price-related information, other indices are needed too. These include the following, some of which are considered in more detail in later sections:

- **The producer price index (PPI)** – this measures price changes of manufactured products from the viewpoint of the manufacturer.

A Practical Introduction to Index Numbers, First Edition. Jeff Ralph, Rob O'Neill and Joe Winton.
© 2015 John Wiley & Sons, Ltd. Published 2015 by John Wiley & Sons, Ltd.
Companion Website: http://www.wiley.com/go/ralph/index_numbers

- **The services producer price index (SPPI)** – this is similar to the PPI, except that it measures the changes in the prices charged by UK businesses for selected services provided to other businesses.

- **Import price index (IPI) and export price index (EPI)** – these measure price movements in goods imported into the UK and exported from the UK, respectively.

Together, this set of price indices – consumer prices indices, producer price indices and the import and export price indices – form a system of price indices [1] and are closely monitored as indicators of the performance of the UK economy.

Other more specific measures are also produced to allow insight into important parts of the economy; these include labour cost indices, which measure changes in the cost of employing labour, the house price index, which is a measure of the movement in house prices and a set of construction price indices, which are used for a variety of purposes, including estimation of construction costs.

Industries where large expenditures are made tend to create specialist price indices. An example is the Ministry of Defence (MOD), whose statisticians produce defence price indices, which capture price movements in the inputs to Defence-related products, including goods, labour and services bought every year. These price indices are used within both the MOD and other Government Departments for financial planning and in writing contracts [2]. Given that prices and the movement of prices are such important economic indicators for so many stakeholders, it is perhaps not so surprising that so many price indices are produced.

11.2   The harmonised index of consumer prices

Although the majority of statistical measures produced in the UK are for UK use, some are produced and supplied to the central, EU-wide statistical organisation – Eurostat, which is a Directorate-General of the European Commission. It provides statistical information on an EU-wide geographical basis. It takes price statistics information from member countries and compiles composite measures for three main geographical groupings:

- Monetary Union Index of Consumer Prices, for Euro area countries

- European Index of Consumer Prices, for EU member countries

- European Economic Area Index of Consumer Prices, for European Economic Area countries.

Contributing countries calculate consumer price inflation according to a common set of standards to produce their individual Harmonised Index of Consumer Prices (HICPs). This is often in addition to their own domestic inflation measures. The specification of common standards is important – without it, the creation of comparable measures across many countries would be much harder and of less value. The development of common, best practice standards and the promotion of these standards is an important function of Eurostat. The methodology of the HICP is specified in a series of legally binding regulations covering the basic construction, and specific regulations for specific aspects of the index[1]; for example, the treatment of insurance. The intention behind the regulations is not to specify the methodology precisely, but rather to ensure that results are comparable while allowing for flexibility in methods. A contributing country can use a different aspect of the methodology as long as it does not lead to difference in results of more than 0.1% in the annual inflation rate [3].

Different types of index are created to facilitate comparisons of price levels across different countries; these are called purchasing power parities (PPPs). These measures are not used for comparison across time, but across geographical areas. That is, they are spatial indices, rather than temporal indices [4].

11.3   UK measures of consumer price inflation

The original measure of consumer price inflation in the UK was the RPI. It was established in 1947 on an interim basis with the first full version published in 1956. The origins of producing an inflation measure in order to adjust wages are much older, originating during the First World War [5]. The RPI was the UK's main measure of inflation until 2002 and is still used for a variety of important purposes, including the indexation of Government Bonds (gilts), the indexation of private pensions and National Savings and Investments index-linking [6].

In addition to the main RPI, a number of variants are produced including RPIX – the RPI excluding mortgage interest payments (MIPs) and the RPIY, which excludes the effect of changes in indirect taxes, such as VAT and excise duties, as well changes in MIPs, housing depreciation and council tax. Collectively, these indices and a number of other variants form the RPI family of price indices.

In 2013, the UK Office for National Statistics carried out a consultation surrounding the use of the Carli formula at the lowest, elementary aggregate level of the RPI. As described in Section 4.2, the Carli formula is an un-weighted arithmetic mean of price relatives. The use of this formula does not meet current international standards and the UK is the only country that continues to use it in their domestic measure [7].[2] Following the consultation, the National Statistician recommended that a new index was required that replaced the Carli with the Jevons; that is, with a geometric mean of price relatives; this formula does meet international standards [8]. The Board of the UK Statistics Authority accepted these recommendations in January 2013. The new index was called the Retail Prices Index Jevons (RPIJ) and was first published in March 2013 [9]. However, the RPI still retains its important functions of indexation of gilts and private pensions; and it also has value as a long-term measure of inflation with only small changes in methodology over time. The comparison of the RPI and

---

[1] There are 21 regulations as of May 2014.
[2] The Carli formula is not permitted within the HICP.

the RPIJ shows the effect of using a different formula at the lowest level of the index.[3] As a result of the RPI not following international best practice, it lost its National Statistics status in March 2013 [10].

The current main measure of inflation in the UK is the CPI. This measure is the UK version of the HICP. It is produced to international best practice standards and achieves a high degree of adherence to the EU HICP regulations. The CPI was first published in the UK in 1996 and became the UK target measure for inflation in 2002, replacing the RPI. Since then, its use has rapidly increased and it is now widely used for a range of purposes including indexing public sector pensions, wages, benefits and tax thresholds [11].

One significant omission from the scope of the index is owner-occupied housing – it was described briefly in Section 8.9; this includes the cost of owning, maintaining and living in one's own house. Housing is a complex good and it has taken many years of study and debate to reach a sufficient consensus within Europe as to how to measure and include it in the CPI; the 'net acquisitions' approach has been adopted as the European standard. There are a number of alternative approaches to measuring changes in the costs of owner-occupied housing and the UK has decided to adopt the 'rental equivalence' approach for its domestic measure, which answers the question: how much would I have to pay in private rent to live a house like mine? The CPI with the inclusion of owner-occupied housing using the rental equivalence measure is called the CPIH that was introduced in March 2013 on an experimental basis [12]. The UK will also produce a version of the CPI including the net acquisitions measure of owner-occupied housing for European comparison purposes.

The UK Office for National Statistics produces a number of other inflation measures, the most important being the gross domestic product (GDP) implied price deflator. It is a broader measure than the CPI in that it covers the whole of the economy and is used to adjust for inflation where a wider measure covering more than just consumer goods is needed. It is produced as part of the UK National Accounts[4] and is derived by comparing the change in the volume of GDP (from which the effect of price changes is removed) with the change in the value of these goods and services.

11.4   PPI and SPPI

It is not just price indices measuring consumer prices that are important economic statistics, price indices for manufactured goods and the provision of services are important too.

The PPI measures the prices of goods that are inputs to manufacturing and prices of outputs from manufacturing [13]. Like the CPI, the construction of this index is based on a basket of representative products; price changes for these products are combined with weights. A sample of manufacturers is approached for price quotes on specific

---

[3] Some parts of the RPI are calculated with the Dutot formula, that is, the ratio of average prices; this formula was not changed in the RPIJ.

[4] https://www.gov.uk/government/publications/gdp-deflators-at-market-prices-and-money-gdp-march-2013.

products. The PPI is considered to be an indicator of emerging inflation pressures and is also used as a deflator in another important output, the Index of Production (IoP).

The Services PPI is similar to the Producer Prices Index; however, as the name suggests, it covers providers of services, which are provided by businesses to businesses;[5] examples include legal and computer services. In total, 43 service industries are included. It is constructed from information collected by a sample survey. Price indices are produced for a variety of service industries but not all as some collections are yet to be developed. They are used in various ways, including for deflation of the Index of Services (IoS) and deflation within the calculation of GDP [14].

11.5   PPPs and international comparison

PPPs are indicators of the difference in price levels between countries [15]. They indicate the relative cost of a particular quantity of goods and services in many countries. They are a form of exchange rate that can be used to make GDP estimates comparable across countries without the distorting effects of differing price levels. They are also used to enable comparisons of productivity across countries.

11.6   Quantity indices

This book concentrates on the construction of price indices, with a particular focus on the measurement of consumer price inflation; the reasons for this choice are explained in Chapter 3. However, in this chapter, we also depart from this focus on price indices and look at indices that measure change in the level of quantities (or volumes) over time rather than change in the level of prices.

Quantity indices are of interest as they measure how much an amount of something has changed over time. This might be the amount of items produced, bought, sold or consumed and can be useful in determining how much a level of activity is changing over a time period, which may be of more fundamental concern than the change in the level of prices. Consider, for example, a shopkeeper who increases his prices by 10% but sees his overall takings fall by 20%; he must be selling less overall than before. In addition, this decrease might not be over the full range of products the shopkeeper sells, and knowing where the differences lie may be important.

In the remainder of this chapter, we take a look at several examples of quantity indices. We will introduce the indices and what they measure and examine particular issues that might affect their estimation; many of these issues are common to measuring change in the general level of prices. These arguments are not discussed here; we focus on the quantity indices most widely used. It should be noted that the choice of index number formula, for example, is just as relevant for the quantity indices presented in this chapter as it is for the price indices considered elsewhere.

---

[5] Services provided to Government Departments are also included.

It is common in both the index number and economic statistics literature to refer to quantity and volume indices interchangeably and that practice is followed in this book as well.

11.7   Gross domestic product

GDP measures total domestic economic activity and is a measure of the size of an economy [16]. This is obviously an extremely complex task and the construction of the measure itself requires understanding of National Accounting Procedures, which are beyond the scope of this book; interested readers are referred to Lequiller and Blades [17]. This framework has been continually developed and improved since its inception with the work of Nobel Prize winning economist Richard Stone.

GDP is a measure of the overall economic activity taking place within a country and is reported quarterly in the UK. The percentage change in quarterly GDP is the headline number, and is an important indicator of the health of the economy, which is watched closely by financial markets. There is extra interest when GDP falls as two quarters of decline is used to define a recession.

The interesting thing about GDP as far as this book is concerned is that it is reported as a quantity (or volume) index. For example, when percentage change in GDP is reported what is really being said is that the index of GDP has increased by that percentage. The time series of the chained volume index for GDP is shown in Figure 11.1. From it, we can see that since the 1950s, the economy of the UK has generally been increasing with occasional dips. Indeed, it is possible to see the most

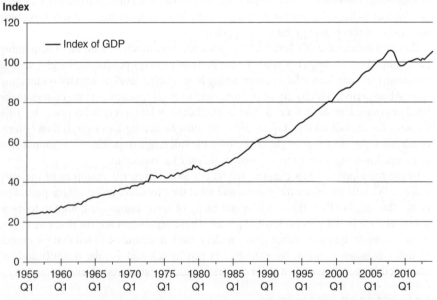

Figure 11.1   Index of GDP, UK, 1955–2013. Source: Office for National Statistics, Series identifier YBEZ.

recent contraction of the UK economy starting in 2009 and that as of 2014 Quarter 1, the UK economy had yet to return to its peak volume achieved in 2008 Quarter 1.

The GDP figures reported in the UK are quarterly estimates. On the day of release, the percentage change in GDP volume index is often the number reported in the news when announced and is speculated on by a range of national and international bodies. Extra importance is afforded to the statistic when the economy has shrunk as two consecutive periods of decline in GDP is commonly used to define the incidence of a recession, which can bring with it all sorts of consequences including additional government actions, as well as affecting how the health of the economy is generally viewed.

GDP is clearly an important economic statistic and this is true internationally with much attention being paid to the GDP measures of countries in the Euro area and the USA as well as the growth of developing countries such as China and India where GDP growth has been more rapid. Some knowledge of index numbers is useful to understand that GDP is at root a volume measure; it also helps those who want to understand key economic statistics.

11.8   Index of Production

The IoP is a quantity index that measures how the 'amount of things' produced in the UK economy has changed over time. Production activities accounted for 15% of the UK economy in 2011 [18] and, therefore, represent an important part of the economy.

Figure 11.2 shows the IoP for the UK and its main sub-indices. These show that production has generally increased since 1948, driven mainly by the Electricity Gas

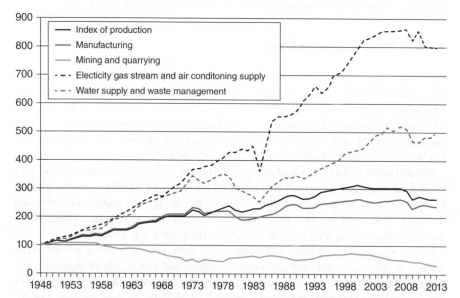

Figure 11.2   Index of production and main sub-indices, UK 1984–2013, rereferenced to 1948 = 100. Source: Office for National Statistics.

Steam and Air Conditioning Supply and Water Supply and Waste Management industries. It also shows the gradual fall in Mining and Quarrying (which includes oil extraction) and shows that manufacturing has fallen slightly over the last 15 years.

The IoP is of obvious interest to economic analysts and policy makers. It is one of the first measures to be released in the quarterly cycle leading up to the announcement of the GDP figure and so is seen as an leading indicator of GDP. The IoP also has a long history and has been reported for several decades, which may also be part of the reason why it has maintained its popularity as a measure of activity in the UK economy while production has declined as a proportion of the whole economy.

The IoP is compiled from surveys of producers, which collect the overall value of the commodities they are producing. This can then be deflated using appropriate deflators, such as the producer price indices discussed in Section 11.4. As a result, deflated measures are obtained in which current period quantities are priced at base period prices. This yields quantity indices for individual areas of production, which compare the current period production to the base period. The form of the quantity index is, therefore, determined by the form of the PPI used. If the deflator used is of a Laspeyres form, then the IoP is a Paasche quantity index and vice versa. However, at a low level within an aggregation structure, the differences between the Laspeyres and Paasche forms of indices are expected to be small. Lower level quantity indices are weighted together to form indices of production at higher levels.

The IoP is a direct input into the GDP calculation and is of interest in its own right. The sub-indices of the IoP are also of interest; for example, if we were to look further into manufacturing, we could look at the indices for food production, production of electrical equipment and the manufacture of chemicals and chemical products. Hence, as with the other indices introduced in this chapter, there is a wealth of information that can help users to understand the UK economy in greater detail.

11.9   Index of services

The economies of developed countries are increasingly focussed on services rather than manufacturing, something which is especially true of the UK where the service sector represented 78% of the economy in 2011 [19]. As a result, it has become increasingly necessary for countries to produce an index of the amount of services consumed within their economy. In this section, we, therefore, briefly describe the UK's IoS; more information is available from ONS [19].

The IoS seeks to measure the changes in the output of the UK service sector, ideally by looking at the growth in the value added by the sector.[6] The IoS is constructed [20] by collecting value measures for the whole range of service sector industries and deflating these using relevant price indices to obtain volume indices. These indices are then weighted together to form the IoS for the overall UK economy.

It is worth noting that these calculations are then used as a proxy for the value added; the difference between the value of outputs and inputs, which represents the value of the service, which is the stated target of the IoS measure [19]. As a result of this, the volume measures are always reported as an index as they serve as a proxy for the real variable of interest.

---

[6] Value added is the increase in the value of goods or services as the result of a production process.

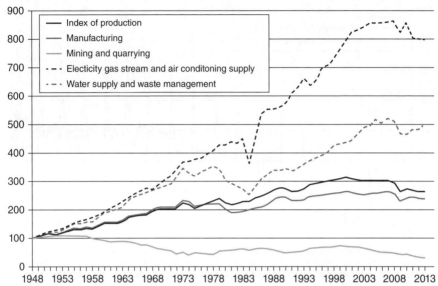

Figure 11.3   Index of services, UK 1997–2013.   Source: Office for National Statistics.

The IoS and its main components are graphed in Figure 11.3, which shows that generally, the service sector in the UK has been growing since 1997; the Real Estate, Scientific, Professional, Technical and Support Activities and the Transport and Storage, Information and Communication Areas have grown quicker than the service sector overall. The IoS is an important input for analysis of the economy.

The IoS is the service sector's equivalent of the IoP. It often receives less attention, possibly as a result of its relatively short lifespan; however, the majority of economic activity in the UK takes place in the service sector; so this index and its associated sub-indices are of increasing importance in analysing and describing changes in the UK economy.

11.10   Retail sales index

How much is being bought by consumers in an economy is an important economic indicator. It is measured in the UK by the Retail Sales Index (RSI) [21], which reports both the value and volume of retail sales in the UK, although in this section, we will focus on the volume of sales.

The index is constructed using a survey of retailers in the UK, with each selected retailer being asked to report on their levels of sales in a variety of categories. Some retailers may thus contribute to several of the sub-indices within the index; for example, a supermarket may contribute information on food retailing, non-food-retailing and automotive fuel. These data are then fed into different low-level indices on retail sales volumes. The sales figures are then deflated using price indices. In the case of the RSI, the deflation is performed using price indices

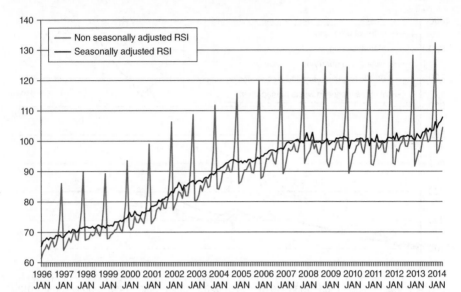

Figure 11.4   Retail sales index adjusted and non adjusted for seasonal effects, UK 1996–2014.   Source: Office for National Statistics.

that are weighted combinations of CPI component price indices where the weights come from the annual business survey, which is also conducted by the ONS [21].

The RSI figures are reported in both non-seasonally adjusted and seasonally adjusted form as are most of the volume indices described in this chapter. Seasonal adjustment techniques are used to remove the effect of regular events, the most obvious of which is Christmas and events that move in time, known as calendar effects, such as Easter [21]. For non-seasonally adjusted retail sales data, it is reasonable to expect a seasonal increase in the volume of retail sales resulting from Christmas shopping. The seasonally adjusted series is presented in Figure 11.4 along with the non-seasonally adjusted series. When using a series, it should be considered whether the raw (non-seasonally adjusted) series is more useful than the seasonally adjusted series.

For the majority of the examples in this chapter, volume indices require price indices in their construction, as well as expenditure shares. The converse is not true in the case of price indices, which do not require volume indices (although expenditure shares are needed for weighting). As a result, quantity indices make greater direct use of the fact that a value index can be decomposed into a quantity and price change element.

11.11   Chapter summary

This chapter has provided a very brief overview of a selection of the most important price and quantity indices that NSIs produce. To produce quantity indices, we usually require price indices as it is not possible to collect the value of fixed quantities from business surveys in a timely or simple manner. For price indices, we can collect the necessary information, which comprises prices and expenditures. To derive quantity indices, NSIs use the equation that shows how value change is made up from the product of a price index and a quantity index as we saw in Sections 4.4, 5.5 and in Chapter 10.

Quantity indices are valuable as indicators of the level of underlying activity in an economy. If we experienced a situation where GDP, IoP, IoS and RSI were all falling, then this would represent a downturn in the economy, something that would not be directly observable from a price index where we can have a rising price level in a growing economy and vice versa. Although these indices are of great use at the aggregate level, however, the lower level versions of these indices also reveal much about the make-up of the fundamental workings of the economy, although more powerful tools than the ones introduced in this book would be needed if we are to look at relationships between multiple indices.

11.12   Data links

ONS Gross Domestic Product Time Series http://www.ons.gov.uk/ons/datasets-and-tables/data-selector.html?dataset=pgdp

ONS IoP Time Series http://www.ons.gov.uk/ons/rel/iop/index-of-production/march-2014/tsd-iop-mar-2014.html

ONS IoS Time Series http://www.ons.gov.uk/ons/datasets-and-tables/data-selector.html?dataset=ios1

ONS Retail Sales Index Time Series http://www.ons.gov.uk/ons/rel/rsi/retail-sales/april-2014/tsd-retail-sales--april-2014.html

References

1. See Chapter 14 in ILO (2004) *Consumer Price Index Manual: Theory and Practice*. The System of Price Statistics, International Labour Office, http://www.ilo.org/public/english/bureau/stat/guides/cpi/#manual (accessed 16 January 2015).

2. Ministry of Defence (2014) Defence Inflation Estimates 2012/2013, https://www.gov.uk/government/uploads/system/uploads/attachment_data/file/312602/ab_defence_inflation_statistical_notice_201213_revised.pdf (accessed 16 January 2015).

3. Eurostat Statistical Books (2008) European Price Statistics – An Overview, p. 27, http://epp.eurostat.ec.europa.eu/portal/page/portal/product_details/publication?p_product_code=KS-70-07-038 (accessed 16 January 2015).

4. Eurostat Purchasing Power Parities, Introduction, http://epp.eurostat.ec.europa.eu/portal/page/portal/product_details/publication?p_product_code=KS-70-07-038 (accessed 16 January 2015).

5. Office for National Statistics (2011) History of and Differences Between the Consumer Prices Index and Retail Prices Index, http://www.ons.gov.uk/ons/rel

/cpi/consumer-price-indices/history-of-and-differences-between-the-consumer-prices -index-and-retail-prices-index/history-of-and-differences-between-the-consumer-price -index-and-retail-price-index---article.pdf (accessed 16 January 2015).

6. Office for National Statistics (2013) Users and Uses of Consumer Price Statistics, http://www.ons.gov.uk/ons/rel/cpi/users-and-uses-of-consumer-price-inflation-statistics /2013/index.html (accessed 16 January 2015).

7. Office for National Statistics (2013) National Statisticians Response to Options for Improving RPI Consultation, http://www.ons.gov.uk/ons/about-ons/get-involved /consultations/archived-consultations/2012/national-statistician-s-consultation-on -options-for-improving-the-retail-prices-index/national-statisticians-response.pdf (accessed 16 January 2015).

8. Office for National Statistics (2013) National Statistician Announces Outcome of Consultation on RPI, http://www.ons.gov.uk/ons/rel/mro/news-release/rpirecommendations /rpinewsrelease.html (accessed 16 January 2015).

9. Office for National Statistics (2013) Introducing the New RPIJ Measure of Consumer Price Inflation, 1997 to 2012, http://www.ons.gov.uk/ons/guide-method/user -guidance/prices/cpi-and-rpi/introducing-the-new-rpij-measure-of-consumer-price -inflation.pdf (accessed 16 January 2015).

10. UKSA (2013) http://www.statisticsauthority.gov.uk/news/statement---retail-prices-index ---14-march-2013.pdf (accessed 16 January 2015).

11. Office for National Statistics (2013). Consumer Price Indices a Brief Guide, http://www .ons.gov.uk/ons/guide-method/user-guidance/prices/cpi-and-rpi/consumer-price-indices --a-brief-guide.pdf (accessed 16 January 2015).

12. Office for National Statistics (2013) Introducing the New CPIH Measure of Consumer Price Inflation, http://www.ons.gov.uk/ons/guide-method/user-guidance/prices/cpi -and-rpi/introducing-the-new-cpih-measure-of-consumer-price-inflation.pdf (accessed 16 January 2015).

13. Office for National Statistics (2014) Producer Price Index, http://www.ons.gov.uk /ons/taxonomy/index.html?nscl=Producer+Price+Indices (accessed 16 January 2015).

14. Office for National Statistics (2014) Services Producer Price Indices, http://www.ons .gov.uk/ons/taxonomy/index.html?nscl=Services+Producer+Price+Indices (accessed 16 January 2015).

15. Eurostat (2014) Purchasing Power Parities, Introduction, http://epp.eurostat.ec.europa.eu /portal/page/portal/purchasing_power_parities/introduction (accessed 16 January 2015).

16. Eurostat (2014) National Accounts and GDP, http://epp.eurostat.ec.europa.eu/statistics _explained/index.php/National_accounts_and_GDP (accessed 16 January 2015).

17. Lequiller, F. and Blades, D. (2006) Understanding National Accounts, OECD, Paris.

18. *Office for National Statistics* (2011) Index of Production Introduction, http://www.ons .gov.uk/ons/guide-method/method-quality/specific/economy/index-of-production/index .html (accessed 16 January 2015).

19. Office for National Statistics (2011) Index of Services Introduction, http://www.ons .gov.uk/ons/guide-method/user-guidance/index-of-services/index.html (accessed 16 January 2015).

20. Office for National Statistics (2011) Methodology of Monthly Index of Services, http://www.ons.gov.uk/ons/guide-method/user-guidance/index-of-services/index-of -services-annex-a--mathematical-formulation-of-the-index.pdf (accessed 16 January 2015).
21. Office for National Statistics (2011) A Quick Guide to the Retail Sales Index, http://www.ons.gov.uk/ons/guide-method/method-quality/specific/economy/retail-sales /a-quick-guide-to-the-retail-sales-index.pdf (accessed 16 January 2015).

12

Further index formulae

12.1 Recap on price index formulae

In Chapter 4, we investigated how to measure price change for a collection of goods and services using the simple, un-weighted price indices of Carli, Dutot and Jevons. Example 4.2 showed that the three different formulae produce different index numbers and, therefore, different measures of inflation. In Chapter 5, two further formulae where introduced, the Laspeyres and Paasche price indices, which include information on quantities as well as prices; the inclusion of quantity information is considered to produce a fairer measure of overall price change. It was shown that the Laspeyres and Paasche formulae can be written as combinations of price relatives and expenditure weights.

Laspeyres and Paasche formulae also yield different index numbers from the same set of prices and quantities as demonstrated by Example 5.1. This leaves producers of consumer price indices with an important decision to make – which formula should be used in practice? Price statisticians who have the responsibility of producing inflation estimates every month need price relatives and expenditure weights in good time. This gives the Laspeyres formula, which uses *base* period expenditure, an advantage over the Paasche formula, which needs *current* weights every month. Chapter 5 explained that the Laspeyres formula is a variant of a more general formula, the Lowe, which allows the weighting information to come from any time period. It is the Lowe formula that price statisticians use in practice, with expenditure weights taken from a time period prior to the base period (i.e. prior to the price reference period). This applies to all levels of the aggregation structure except at the lowest level, where, in most cases, weighting information is not available and simple un-weighted formulae are used, as explained in Section 8.7.

A Practical Introduction to Index Numbers, First Edition. Jeff Ralph, Rob O'Neill and Joe Winton.
© 2015 John Wiley & Sons, Ltd. Published 2015 by John Wiley & Sons, Ltd.
Companion Website: http://www.wiley.com/go/ralph/index_numbers

12.2   Classifying price and quantity index formulae

The Laspeyres and Paasche price and quantity indices are not the only functional forms that can be used to combine price and quantity (or price relative and expenditure weight) information to produce overall price and quantity index numbers. There are very many other possibilities, and we will explore just a few of them in this chapter; more comprehensive reviews can be found in the books of Balk and von der Lippe [1].

Price indices are mathematical functions that combine price and sometimes quantity information to form price index numbers. They fall into two broad types, weighted and un-weighted. The weighted indices can be further classified into those that use weighting information from one time period – 'asymmetrically weighted indices' and those that use weighting information from two periods in an equal way – 'symmetrically weighted indices'.

The un-weighted formulae are some of the earliest proposed ways of combining price information and it might be thought that they are only of historical value. However, as shown in Chapter 8, they are still in use today, appearing at the lowest level of a consumer price index where weighting information is not usually available. Choosing which of the alternative un-weighted formulae to use is still a controversial issue.

In this chapter, we will only consider price indices; however, for every formula, there is a corresponding quantity index that can be obtained easily by substituting quantity relatives for price relatives.

12.3   Asymmetrically weighted price indices

The most well known of price indices that use expenditure weights from only one period are the price index formulae of the German economists, Étienne Laspeyres and Hermann Paasche, which were introduced in Chapter 5:[1]

$$P_{\text{Laspeyres}} = \frac{\sum_i p_{ti} q_{0i}}{\sum_i p_{0i} q_{0i}} = \sum_i \left(\frac{p_{ti}}{p_{0i}} \right) . s_{0i} \tag{12.1}$$

$$P_{\text{Paasche}} = \frac{\sum_i p_{ti} q_{ti}}{\sum_i p_{0i} q_{ti}} = \frac{1}{\sum_i s_{ti} \left(\frac{p_{ti}}{p_{0i}} \right)^{-1}} \tag{12.2}$$

They are shown in both the 'price, quantity' form and the 'price relative, expenditure weight' form. The expenditure weights are given by:

$$s_{0i} = \frac{p_{0i} q_{0i}}{\sum_j p_{0j} q_{0j}} \quad \text{and} \quad s_{ti} = \frac{p_{ti} q_{ti}}{\sum_j p_{tj} q_{tj}} \tag{12.3}$$

In the 'price relative, expenditure weight' form, the Laspeyres price index is a 'base-period expenditure weighted, arithmetic mean of price relatives' and the

---

[1] For this chapter and the next, we have changed the style, putting the time period as a subscript; this is a neater representation when many formulae appear. A discussion of styles can be found in Appendix A, section 3.

Paasche Price index is a 'current period, expenditure weighted, harmonic mean[2] of price relatives'.

It is interesting to note that although a variation of the Laspeyres formula is used by almost all National Statistics Institutes to produce their consumer price indices, Étienne Laspeyres thought that his formula was of no practical use as quantity information would not be readily available. It was the economist Irving Fisher who showed how it could be written in the 'price relative, expenditure weight' form and thus it became a practical way of measuring price change [2].

Two further formulae can be created by adjusting the formulae used in the Laspeyres and Paasche indices. We can achieve this by swapping the combinations of the weighting period and the type of mean of price relatives. If we combine current period weights with an arithmetic mean of price relatives, we get the Palgrave Index [3]:

$$P_{\text{Palgrave}} = \sum_i s_{ti} \left(\frac{p_{ti}}{p_{0i}} \right) \tag{12.4}$$

In a similar way, if we use base-period weights with a harmonic mean of price relatives, we obtain the harmonic Laspeyres price index:

$$P_{Harmonic\ Laspeyres} = \frac{1}{\sum_i s_{0i} \left(\frac{p_{ti}}{p_{0i}} \right)^{-1}} \tag{12.5}$$

There are two further variants of Laspeyres and Paasche formulae that are important to consider which use geometric, rather than arithmetic means of price relatives. The geometric Laspeyres is a base-period, expenditure weighted, geometric mean of price relatives and the geometric Paasche is a current-period expenditure weighted geometric mean of price relatives; they can be written algebraically as:

$$P_{G,\text{Laspeyres}} = \prod_i \left(\frac{p_{ti}}{p_{0i}} \right)^{s_{0i}}$$

$$P_{G,\text{Paasche}} = \prod_i \left(\frac{p_{ti}}{p_{0i}} \right)^{s_{ti}} \tag{12.6}$$

Equations (12.1–12.6) consist of the six possible combinations of base and current period weights (two types of weighting) and the arithmetic, geometric and harmonic means of price relatives (three types of mean).

It is possible to both combine and extend these six formulae into one, generalised mean formula:

$$P_{\text{GM}}(r) = \left[\sum_{i=1}^{N} \bar{s}_i \left(\frac{p_{ti}}{p_{0i}} \right)^r \right]^{1/r} \tag{12.7}$$

---

[2] The harmonic mean of a series of values x_i, where $i = 1, \ldots, n$ is $n / \sum_{i=1}^{n}(1/x_i)$.

If we set the weights $\bar{s}_i$ and the power, r, as follows,[3] then the generalised mean reduces to the six previously defined formulae:

Laspeyres $\bar{s}_i = s_{0i}$ $r = 1$
Paasche $\bar{s}_i = s_{ti}$ $r = -1$
Palgrave $\bar{s}_i = s_{ti}$ $r = 1$
Harmonic Laspeyres $\bar{s}_i = s_{0i}$ $r = -1$
Geometric Laspeyres $\bar{s}_i = s_{0i}$ $r = 0$
Geometric Paasche $\bar{s}_i = s_{ti}$ $r = 0$.

The correspondence between the generalised mean and the first four formulae above is straightforward to see. For the last two, the geometric versions need careful mathematical analysis.

The formulae mentioned above have introduced some additional ways of combining price and quantity information in order to produce a price index to measure change in the general level of prices. In fact, there are many other possible formulae [4]; further indices are given by Balk [5] and von der Lippe [6].

We now turn to a general class of indices, which are called basket indices. A basket index is one of the form:

$$P = \frac{\sum_i p_{ti} q_i}{\sum_i p_{0i} q_i} \tag{12.8}$$

That is, the ratio of the prices of a basket of goods where the quantities, q_i, can refer to any time period, or a combination of time periods [7].

If we set the quantities to refer to the base period, t_0, then it is easy to see that formula 12.8 mentioned above reduces to the Laspeyres price index (Equation 12.1). Alternatively, if we set the quantities to refer to the current, or comparison period, then we arrive at the Paasche formula (Equation 12.2).

The general form of the basket index with the time period set to an arbitrary time period, labelled b, is known as the Lowe formula – this was introduced in Section 5.7. The formula is:

$$P_{\text{Lowe}} = \frac{\sum_i p_{ti} q_{bi}}{\sum_i p_{0i} q_{bi}} \tag{12.9}$$

It is possible to write the Lowe price index in the 'price relative/expenditure weight' form with a little algebra:

$$P_{\text{Lowe}} = \frac{\sum_{i=1}^{N} p_{ti} q_{bi}}{\sum_{i=1}^{N} p_{i0} q_{bi}} = \frac{\sum_{i=1}^{N} p_{ti} q_{bi} \left[\frac{p_{0i}}{p_{0i}} \frac{p_{bi}}{p_{bi}} \right]}{\sum_{i=1}^{N} p_{i0} q_{bi} \left[\frac{p_{bi}}{p_{bi}} \right]} = \frac{\sum_{i=1}^{N} \left[p_{bi} q_{bi} \frac{p_{0i}}{p_{bi}} \right] \frac{p_{ti}}{p_{0i}}}{\sum_{i=1}^{N} \left[p_{bi} q_{bi} \frac{p_{0i}}{p_{bi}} \right]}$$

---

[3] Note that as r could theoretically take any values, we could claim that there are an infinite number of possible index formulae within the definition of the generalised mean.

$$= \sum_i s'_{bi} \cdot \left(\frac{p_{ti}}{p_{0i}} \right),$$

$$\text{with } s'_{bi} = \frac{p_{bi} q_{bi} \left[\frac{p_{0i}}{p_{bi}} \right]}{\sum_j p_{bj} q_{bj} \left[\frac{p_{0j}}{p_{bj}} \right]} \tag{12.10}$$

The weights consist of the period b expenditure shares, but multiplied by the ratio of prices at time 0 to b. In practice, the time period b occurs before time period 0, and the factor is sometimes described as 'price updating the weights'. The Lowe index is particularly important as it is the formula that almost all National Statistical Institutes use in for their consumer price indices. As noted in Chapter 8, the time period b, which is the weight reference period, is typically a year in duration and is before the time period 0, the price reference period; both time periods 0 and t are a month.

In the literature, the Lowe Index is sometimes described as a 'Laspeyres-type' Index; however, one could argue that a more accurate description is that the Laspeyres is a 'Lowe-type' index.

A slightly different formula incorporating a weight reference period b is attributable to Young [8], an English economist:

$$P_{\text{Young}} = \sum_i \left(\frac{p_{ti}}{p_{0i}} \right) \cdot s_{bi}, \text{ with } s_{bi} = \frac{p_{bi} q_{bi}}{\sum_j p_{bj} q_{bj}} \tag{12.11}$$

This has the weights as period b expenditure shares. There are also geometric versions of the Lowe and Young indices; the geometric Young index is also known as the Cobb-Douglas index, which is discussed in Section 13.3.

12.4   Symmetric weighted price indices

From a theoretical viewpoint, that is, leaving aside the practical concerns about the availability of weights, both the Laspeyres and the Paasche indices use equally valid, natural choices of weight reference period. At the end of the nineteenth century and in the early twentieth century, economists proposed a compromise between the Laspeyres and the Paasche formulae, which uses weighting information from both the base and comparison periods, yielding two 'symmetric' formulae. It was expected that such formulae would be an improvement over 'asymmetric' forms, which only use weighting information from one time period. In practice, the prices of goods and services change between the two time periods 0 and t; this results in consumers adjusting the quantities they buy, and is reflected in the difference in weights in the two time periods. By incorporating weights from the two time periods, it was thought that some aspects of this consumer behaviour would be captured.

There are two ways to combine weights from the two time periods in a symmetric way – an arithmetic mean of quantities or a geometric mean. These averaged quantities are combined with prices in an arithmetic sum.

The Marshall-Edgeworth formula takes the *arithmetic* mean of the quantities in the two time periods:

$$P_{\text{Edgeworth-Marshall}} = \frac{\sum_i p_{ti}\cdot\left(\frac{q_{0i}+q_{ti}}{2}\right)}{\sum_i p_{0i}\cdot\left(\frac{q_{0i}+q_{ti}}{2}\right)} \tag{12.12}$$

The Walsh formula uses the *geometric* mean of the quantities in the two time periods:

$$P_{\text{Walsh}} = \frac{\sum_i p_{ti}\cdot\sqrt{q_{0i}\cdot q_{ti}}}{\sum_i p_{0i}\cdot\sqrt{q_{0i}\cdot q_{ti}}} \tag{12.13}$$

A further formula arises from a combination of an arithmetic mean of weights with a geometric combination of price relatives. This gives the Törnqvist formula:

$$P_{\text{Tornqvist}} = \prod_i \left(\frac{p_{ti}}{p_{0i}}\right)^{(1/2).(s_{0i}+s_{ti})} \tag{12.14}$$

In this formula, the price index takes the geometric mean of the price relatives, each raised to the power of the arithmetic average of the expenditure shares of the two time periods.

An alternative approach to reconciling the Laspeyres and Paasche formulae is to go further than combining the weights and to combine whole formulae; again there are two options, an arithmetic combination and a geometric one.

Drobisch proposed the arithmetic mean of the Laspeyres and Paasche formulae:

$$P_{\text{Drobisch}} = \frac{1}{2}.(P_{\text{Laspeyres}} + P_{\text{Paasche}}) \tag{12.15}$$

and Fisher [4] proposed the geometric mean of the Laspeyres and Paasche formulae:

$$P_{\text{Fisher}} = \sqrt{P_{\text{Laspeyres}}.P_{\text{Paasche}}} \tag{12.16}$$

Irving Fisher called his index 'ideal' as it satisfied a number of attractive mathematical properties, which will be described in Chapter 13.

12.5   Un-weighted price indices

All the previous formulae in this chapter use weighting information. There is a class of indices that only contain price information; these indices do not include information about quantities and, therefore, weights. They can be said to be less desirable than the weighted formulas as they include price changes of all commodities equally, taking no account of the relative importance (via an expenditure weight) of commodities. They were introduced in Chapter 4.

It was noted in Chapter 8 that compilers of consumer prices index numbers would prefer to only use weighted formulae; however, weighting information is not available at the elementary aggregate level, so un-weighted formulae are used and constitute an important building block of the overall index. The question of which formula is 'best' at this level is still a controversial matter[4]. The three, well-known, simple or un-weighted price index formulae are the Dutot, the Carli and the Jevons.

The three main un-weighted formulae were introduced in Chapter 4; we just summarise their mathematical forms here:

The Dutot formula – ratio of average prices:

$$P_{\text{Dutot}} = \frac{\sum_i p_{ti}}{\sum_i p_{0i}} \tag{12.17}$$

The Carli formula – the average of price relatives:

$$P_{\text{Carli}} = \frac{1}{N} \sum_i \frac{p_{ti}}{p_{0i}} \tag{12.18}$$

The Jevons formula – geometric average of price relatives:

$$P_{\text{Jevons}} = \left(\prod_i \frac{p_{ti}}{p_{0i}} \right)^{1/N} \tag{12.19}$$

There are two further formulae that are worth noting at this point – the harmonic mean of price relatives:

$$P_{Harmonic} = \left[\frac{1}{N} \sum_i \frac{p_{ti}}{p_{0i}} \right]^{-1} \tag{12.20}$$

and the Carruthers, Sellwood, Warn and Dalen Index, which is the geometric mean of the Carli and the Harmonic mean of price relatives:

$$P_{\text{CSWD}} = \sqrt{P_{\text{Carli}} \cdot P_{\text{Harmonic}}} \tag{12.21}$$

Although little used, this is an interesting formula as it is the un-weighted index that corresponds to the (weighted) Fisher index. In fact, all of the un-weighted formulas in this section have weighted 'equivalents'; for example, the Carli corresponds to the Laspeyres and the Jevons to the geometric Laspeyres.

---

[4] See, for example, the responses to the National Statistician's consultation on improving the RPI at http://www.ons.gov.uk/ons/about-ons/get-involved/consultations/archived-consultations/2012/national-statistician-s-consultation-on-options-for-improving-the-retail-prices-index/responses-to-options.zip

12.6   Different formulae, different index numbers

This chapter has described just some of the many formulae that have been proposed as measures of price change for a set of commodities. A key question is whether these different formulae lead to different index numbers and so to different values for inflation. If they all give us measures of price change which are very close, then the choice of formula is a matter of academic interest only. This section will use a sample data set to examine differences that result from using different formulae; these differences should make it clear that the choice of formula can have an impact on the change in the price level we are seeking to measure.

Consider Table 12.1 of prices and quantities for four types of loose tea over three time periods.[5] Using these data, we can calculate index numbers with a selection of index formulae (Table 12.2).

Note that the index numbers differ between the different formulae. In the table, the order has been chosen so that the index numbers calculated are shown in descending order (Figure 12.1).

This simple example shows that different index formulae can produce different index numbers. These differences extend to the values of inflation calculated from these index numbers. This means that a choice of formula is required so that one definitive set of index numbers can be produced. The differences in the example might

Table 12.1   Price and quantity data for sales of tea.

Fine tea (100 g)	2010		2011		2012		2013	
	p_0	q_0	p_1	q_1	p_2	q_2	p_3	q_3
Oolong	6.3	34	6.5	28	6.9	27	7.2	28
Darjeeling	5.9	37	6.8	31	7.2	27	7.8	22
Lapsang Souchong	5.8	44	5.5	55	5.4	58	5.7	53
Jasmine	6.1	65	6.8	58	7.3	57	7.7	50

Table 12.2   Index numbers for sales of tea using a variety of index formulae.

	2011	2012	2013
Laspeyres	106.68	111.89	118.49
Geometric Laspeyres	106.39	111.28	117.78
Törnqvist	106.14	110.75	116.99
Fisher	106.13	110.70	116.95
Geometric Paasche	105.89	110.22	116.21
Paasche	105.58	109.52	115.43

---

[5] These are not real prices and quantities; they are just for illustration.

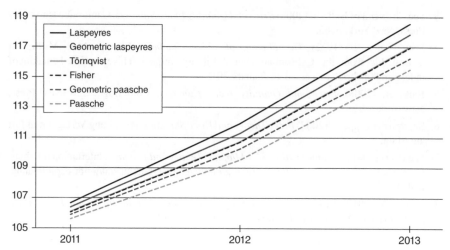

Figure 12.1   Index numbers from different formulae with the same price and quantity data.

seem small at first glance; however, remember that even a small difference in the value of inflation can have a big impact on the way an economy is viewed.

12.7   Chapter summary

This chapter has described some of the formulae that have been proposed as measures of price change for a set of commodities; there are many others. The basic question is how to produce an average of a set of changing prices and quantities, and there are very many ways to do this. The fact that the different ways of doing this produce materially different index numbers and values of inflation is a problem, in fact, it is a very great problem. Given the importance of the measurement of the general level of prices, much theoretical and practical work has gone into trying to resolve the key question: which is the best formula to use?

We noted in Chapters 4 and 8 that it is the Lowe formula that is used by almost all National Statistics Institutes in their monthly calculation of the rate of inflation. Does this mean that it is the best formula when compared with the alternatives presented in this chapter? This is a question that will be explored in Chapter 13.

References

1. See Chapters 3, 5 and 6 of Balk, B. (2008) *Price and Quantity Index Numbers*, Cambridge University Press, New York, and Chapters 2 and 3 of von der Lippe, P. (2007) *Index Theory and Price Statistics*, Peter Lang Verlag, Frankfurt am Main..

2. See Chapter 1 of Balk, B. (2008) *Price and Quantity Index Numbers*, Cambridge University Press, New York, p. 7.

3. See Chapter 3 of Balk, B. (2008) *Price and Quantity Index Numbers*, Cambridge University Press, New York, p. 63.

4. See: Fisher, I. (1922) *The Making of Index Numbers*. Boston, MA: Mifflin. It is available from the Californian Digital Library, https://archive.org/details/makingof indexnum00fishrich (accessed 16 January 2015).

5. Balk, B. (2008) *Price and Quantity Index Numbers*, Cambridge University Press, New York.

6. See von der Lippe, P. (2007) *Index Theory and Price Statistics*, Peter Lang Verlag, Frankfurt am Main.

7. See ILO (2004) *Consumer Price Index Manual: Theory and Practice*, International Labour Organisation, Geneva, http://www.ilo.org/public/english/bureau/stat/guides/cpi/#manual p. 451.

8. Young, A. (1812) *An Inquiry into the Progressive Value of Money in England as Marked by the Price of Agricultural Products*, Hatchard, Piccadilly.

Exercise J

Solutions to this exercise can be found in Appendix D.

J.1 A basket index takes the following form:

$$P_{\text{Basket}} = \frac{\sum_{i=1}^{N} p_{ti} q_i}{\sum_{i=1}^{N} p_{0i} q_i}$$

a. Show that it can be written in the 'weight/price relative' form:

$$P_{\text{Basket}} = \sum_{i=1}^{N} w \cdot \left(\frac{p_{ti}}{p_{0i}}\right)$$

Describe the weights.

b. Show that the Lowe price index can be written as a ratio of two Laspeyres price indices:

$$P_{\text{Lowe}} = \frac{\sum_{i=1}^{N} p_{ti} q_{bi}}{\sum_{i=1}^{N} p_{0i} q_{bi}} = \frac{P_L^{b,t}}{P_L^{0,b}}$$

c. Show that the Törnqvist price index is the geometric mean of the geometric Laspeyres price index and the geometric Paasche price index.

d. A price index satisfies the time reversal property if the following is true:

$$P^{0,t} \cdot P^{t,0} = 1$$

Show that the (arithmetic) Laspeyres price index fails this test, while the Törnqvist price index passes.

J.2 Consider the following price and quantity data (Table 12.3):

a. Calculate the (arithmetic) Laspeyres and Paasche price index numbers for this data where time period '1' is the comparison period and time period '0' is the base period.

Table 12.3 Price and quantity data.

Item	p_{0i}	q_{0i}	p_{1i}	q_{1i}
1	2.1	12	2.6	10
2	2.4	9	2.7	7
3	2.9	14	2.8	16
4	3.0	10	3.4	9

b. Calculate the Geometric Laspeyres price index number and the Geometric Paasche price index number.

c. i. Calculate the Fisher price index number using the fact that it is the geometric mean of the (arithmetic) Laspeyres and Paasche price index numbers.

 ii. Calculate the Törnqvist price index number using the fact that it is the geometric mean of the geometric Laspeyres price index number and the geometric Paasche index number.

d. For the index numbers calculated in part c, calculate the following:

 i. The percentage difference between the Fisher and the Törnqvist index number

 ii. The percentage difference between the geometric Laspeyres and geometric Paasche index numbers

 iii. The (arithmetic) Laspeyres and Paasche price index number.

 iv. Describe the results of parts i–iv.

13

The choice of index formula

The question of what is the 'best' formula to measure the change in the general level of prices has been studied for many years; it has been and remains a contentious issue. It is a difficult question to answer and in attempting to answer it, index theorists and price statisticians have explored four different approaches – the axiomatic, the economic, the sampling and the stochastic.

This chapter starts by recalling the index number problem, before looking briefly at the four approaches. A more technical treatment of the axiomatic and economic approaches is given in Appendix B. The chapter then considers the conclusions of these theoretical studies of index formulae and examines how practical considerations have influenced the choice of formula. The chapter concludes with a summary.

13.1   The index number problem

The index number problem was discussed earlier in Sections 4.4 and 5.5; it is the problem of decomposing changes in an aggregate amount of money into overall price and quantity changes. Consider the shopping basket described in Table 13.1 for an individual consumer's weekly purchase of fruit. The prices change across the two weeks and so does the amount of each item bought. In total, in week 1, £3.75 was spent on fruit; in the second week, the expenditure was £3.15. If we look at the change in spending in the second week (using the first week as a base) then we see a 16% decrease in spending on fruit over the week. The index number problem occurs when we want to split this change in total amount spent (or total amount received if we are looking at this from the perspective of the fruit seller) into changes caused by the changes in prices and changes caused by the differences in quantities consumed

A Practical Introduction to Index Numbers, First Edition. Jeff Ralph, Rob O'Neill and Joe Winton.
© 2015 John Wiley & Sons, Ltd. Published 2015 by John Wiley & Sons, Ltd.
Companion Website: http://www.wiley.com/go/ralph/index_numbers

Table 13.1   Fruit prices and quantities.

Fruit	Week 1		Week 2	
	Price	Quantity	Price	Quantity
Apples	£0.50	2	£0.45	3
Oranges	£0.25	4	£0.30	2
Bananas	£0.35	5	£0.40	3

in the two weeks. That is, can we come up with a pair of quantity and price indices which satisfy the following formula?

$$Value^{0,t} = Price^{0,t} \times Quantity^{0,t}$$

where $Price^{0,t}$ is a price index between periods 0 and t, and $Quantity^{0,t}$ is a quantity index between those two periods.

The split of the change in value above can be used to determine which price and quantity indices should be used together if we wish to ensure that changes in an overall amount can be split into changes in price levels and changes in quantity levels. We saw in Chapter 10 that the combination of a Laspeyres price index and a Paasche quantity index preserves the above identity as does the combination of a Paasche price index with a Laspeyres quantity index. In fact, a Fisher quantity index coupled with a Fisher price index also decomposes the overall change in value.[1]

Using the data in Table 13.1, we find the following results (to four decimal places):

$$V^{1,2} = 0.8400$$

$$P_L^{1,2} = 1.0933, \quad Q_P^{1,2} = 0.7683$$

$$P_P^{1,2} = 1.0328, \quad Q_L^{1,2} = 0.8133$$

$$P_F^{1,2} = 1.0626, \quad Q_F^{1,2} = 0.7905$$

As was shown in Chapter 10, this knowledge can be useful when determining the appropriate index to use; for example, Statistics Canada produces International Merchandise Trade price and volume indices. In order to produce a Laspeyres volume index, a value ratio is divided by a Paasche price index [1].

The index number formula is an important part of understanding the role of index numbers as it is the first point at which we have considered how index numbers might behave when combined rather than thinking about them in isolation as measuring a change in a single variable. Placing them in this context can be important when we think about their role in measuring changes in total expenditures or revenues. If we choose a target formula, for example, a *Laspeyres* quantity index, we use the decomposition of value change to identify an appropriate companion

---

[1] These statements are proven in Appendix A; however, interested readers might attempt to prove these results for themselves.

formula. Notice that decomposition does not yield a single pair of indices – there are multiple solutions.

13.2   The axiomatic approach

The axiomatic, or test approach to index numbers, examines the mathematical properties of indices.[2] A number of desirable properties have been proposed which can be used to assess the quality of an index formula; the axiomatic approach to index numbers therefore involves evaluating the properties of the functional forms of the index. The aim is to identify those indices which fulfil many of these properties; the ideal position being that a single index can be found that satisfies all desirable properties. This approach was made famous in the index number world by Irving Fisher's monumental work on index numbers – The Making of Index Numbers [2]. In this book Fisher looked at the properties of over 200 indices and variants and concluded that the one which best fulfilled these desired properties was the index which has borne his name ever since.

Not all of the proposed properties are considered as being equally important. Some are considered to be self-evident properties that any valid index should have (they are 'axiomatic'), while others are important, but not essential. Some of the proposed properties are such that few people would dispute them; for example, if the price of every good increases then it seems sensible that a statistic that measures changes in the level of prices should also increase. Others are more debatable; for example, there has been much discussion in the index number community about whether an index number formula should fulfil what is known as the time reversal property. The time reversal property requires that if a set of prices is observed in periods 0 and t and we use an index to calculate:

1. an index number of price change from period 0 to period t ($= P_{0t}$)

2. an index number of price change from period t to period 0 ($= P_{t0}$).

Then: $P_{0t} = P_{t0}^{-1}$

The Jevons formula (or geometric mean) is fulfils this property, but the Carli index does not. Some who advocate the axiomatic approach would argue that this is a fatal flaw of the Carli index, while others would argue that the time reversal property is not axiomatic and that we can do without it for our index formulae.

The debate around the time reversal property illustrates the central problem with the axiomatic approach: different properties are axiomatic for different people. For every index, it is possible to identify a property that it will not pass. Efforts have been made to identify a set of properties that most index number experts agree on [3], but while there is still debate around a set of undisputed axioms and tests, the approach cannot be said to have identified a definitive, ideal index.

---

[2] Remember that an index is a mathematical function while an index number is a number.

13.3   The economic approach

The economic approach to index numbers uses ideas common within economic theory to identify what should be measured by an index number to determine changes in the level of prices. Economic theory concerns itself less with averaging changes in individual prices than with measuring the change in overall income required for individuals to maintain a constant standard of living, or more precisely in economic terms to maintain a constant level of utility. Economic theory proposes that there are many combinations of products which can yield a given level of utility and if consumers are behaving in a rational way they will select the combination of goods which achieves a given level of utility at the lowest cost. Therefore when prices change, economists will require a price index to measure the change in cost to achieve a given level of utility – this is known as a 'cost of living' index. This idea was first introduced by Bennet [4], Bowley [5] and more formally by Konüs [6].

In order to make matters tractable, economists employ the idea of a cost function $C(p^t, u)$ where the function C takes as inputs a vector[3] of prices at time t, $p^t = (p_1, \ldots p_N)$, and given level of utility, u, and the function produces the minimum cost of achieving the stated utility level. A Konüs, or economic price index, between periods 0 and t, is therefore:

$$\frac{C(p^t, u)}{C(p^0, u)} \tag{13.1}$$

An issue arises in that we have taken the utility level used in the index as being given; however, as with the magnitudes of the quantities used in more traditional indices there are several choices. We might use the utility achieved in the original period, u_0, to give a Laspeyres–Konüs index, or the utility achieved in period t, u_t, to give the Paasche–Konüs index or some other amount; for example, one which represents an average or minimum utility level. Hence we still have choices to make in terms of which utility level to use.

It is possible to identify certain specialised results regarding economic indices. A Jevons index can be shown to be a Konüs index in the special case where the consumer's utility function is represented by specific form of the Cobb–Douglas utility function [7]. This result has been used to suggest that the Jevons index measures changes in consumer substitution [8]. That the Jevons is a good Konüs index in this specified theoretical setting does not mean that it should be immediately adopted as an economic price index in practice, unless the assumptions required to make this argument are also fulfilled, for which, at the moment, there is insufficient evidence. Although the economic approach provides a framework with which to investigate the suitability of different indices, it is not easy to apply in practice.

A further issue is that we have taken the form of the cost function for granted. In fact, the function is assumed to have several properties without the exact form of such a function being known, making it very difficult for price statisticians to measure cost of living indices. The work of Diewert [9] investigates how well certain classes

---

[3] This is a collection of prices for N goods and services.

of indices can approximate an economic index and concludes that the Fisher index does this well if a number of assumptions are made. Other suggestions for measuring a Konüs type index include using a technique called demand system analysis [10]. In conclusion, although the theory of the cost of living index is well understood, putting the approach into practice has proven more difficult. As more detailed data and research on consumer behaviour become available we may be able to measure an economic index reliably in the future.

13.4   The sampling approach

If we think of the measure of the change in the level of prices as a population statistic; that is, a characteristic of a population of prices, it is possible for us to frame the problem of how to estimate it from a sample of prices. Thus we consider how a sample based estimate might be constructed to best estimate the population statistic of interest [11]. This approach might begin by posing the central question of how we would construct a measure of inflation if we had access to price and quantity information for every transaction in the economy.

The sampling approach selects an estimator and sampling method which will provide the best estimate of the population parameter in terms of accuracy and bias. This becomes more difficult when we have to use un-weighted indices at the elementary aggregate level in order to estimate weighted indices. When weighting together sub-indices we might also have to think about whether this maintains the right form of the index. For example, a weighted combination of Laspeyres indices is itself a Laspeyres index; however, this is not true for a Fisher index. The idea behind the sampling approach is therefore to decide on the most accurate estimator of a given population measure of inflation.

As with the other approaches described in this chapter there are issues; the population statistic is rarely defined and never measured; the sampling design is often inherited and developed asymmetrically across categories. Despite this, the sampling approach can be a useful way to think about choosing an index and if we were creating a measure of inflation from scratch might be a useful way to proceed, albeit after the difficult first step of choosing the target population index. Alternatively, one could evaluate current approaches to price indices by using their sampling methodology and sample estimator to infer what the target index is. For example, if we were taking weighted combinations of un-weighted Carli indices this might imply that our target index was a Laspeyres index. For more complex designs this will be a difficult task in itself but may provide some information about what an index is trying to achieve.

An interesting note on the sampling approach is that as scanner data, which is transaction level data stored electronically, becomes available to price statisticians it provides opportunities for researchers to simulate and test different sampling approaches and estimators. For example, it could prove useful for investigating which un-weighted elementary aggregate can best approximate a weighted index.

13.5   The stochastic approach to index numbers

A limitation of the axiomatic and the economic approaches to index numbers is that they cannot provide a measure of the accuracy of the index numbers that result from using the selected formulae. In contrast, the stochastic approach, which is based on models of inflation, produces measures of accuracy as an integral part of the method.

The stochastic approach to index numbers treats the price movement of a commodity as consisting of an underlying, common inflation component together with a random variation. A simple application of this approach is the following model:

$$\frac{p_{ti}}{p_{0i}} = \alpha_{0t} + \varepsilon_{0ti} \quad i = 1, 2, \dots, n \tag{13.2}$$

where α_{0t} is the common inflation rate across the collection of commodities and the ε_{0ti} are independent random variables with mean of zero and variance σ^2. Given a set of price data, we can use statistical analysis to determine an unbiased estimator of the parameter α_{0t}. For this model, the least squares or maximum likelihood estimator for α is the Carli formula; that is, the arithmetic mean of the price relatives:

$$\hat{\alpha}_{0t} = \frac{1}{n} \sum_{i=1}^{n} \frac{p_{ti}}{p_{0i}} = P_C^{0,t} \tag{13.3}$$

A different assumption can be made for the model to describe inflation:

$$\ln\left(\frac{p_{ti}}{p_{0i}}\right) = \beta_{0t} + \varepsilon_{0ti} \quad i = 1, 2, \dots, n \tag{13.4}$$

where: $\beta = \ln \alpha$ and the ε_i are random variables as described above. The least squares estimator of β is the arithmetic mean of the *logarithm* of the price relatives:

$$\hat{\beta}_{0t} = \frac{1}{n} \sum_{i=1}^{n} \ln\left(\frac{p_{ti}}{p_{0i}}\right) \tag{13.5}$$

that is, the geometric mean of the price relatives; hence the common inflation rate α_{0t}, is estimated by the Jevons formula:

$$\hat{\alpha}_{0t} = P_J^{0,1} = \left(\prod_{i=1}^{n} \frac{p_{1i}}{p_{0i}}\right)^{\frac{1}{n}} \tag{13.6}$$

Although the parameter $\hat{\beta}_{0t}$ is an unbiased estimator of β_{0t}, the subsequent estimate of the common inflation rate given by the Jevons formula, found from taking the exponential of both sides of Eq. 13.4, is not; it has an upward bias [12].

The stochastic approach has been criticised by economists as not being supported by the evidence of price movements; that is, models fit poorly [13]. Prices of different goods and services move differently and therefore do not have a common underlying inflation rate as required by the two models above. This criticism can be overcome, to a degree, by limiting the range of goods for which price quotes are aggregated to comparable commodities which have similar price movements.

The question of which of the two models of an underlying, common inflation parameter to choose can be explored by looking at the distributions of price relatives incorporated into the Consumer Prices Index. If our target is a measure of central tendency, then the geometric mean is an appropriate measure if the underlying probability distribution function is log-normal. Alternatively, if the distribution of price relatives is normally distributed then the arithmetic mean is the appropriate measure of central tendency. The shape of price relative distributions has been examined for the UK Consumer Prices Index [14]. Statistical tests show that the actual distributions fit neither lognormal nor normal distributions very well, though about two thirds of the distributions were found to be closer to a lognormal distribution than a normal one. This could be considered as supporting a slight preference for the Jevons formula at the elementary aggregate level.

The models can be extended and all of the familiar index formulae (including weighted formulae) can be derived from specific regression models [15]. With a specific set of price data, it is possible to use statistical criteria to evaluate the 'best' model; however, this is not the same as saying the index formula is 'best' in all applications to price data.

13.6   Which approach is used in practice?

We have briefly examined four approaches to index numbers. The two major approaches – the axiomatic and the economic and the two minor approaches – the sampling and the stochastic. What should we conclude? Although there is still much debate on the subject, there is a significant body of expert opinion that favours superlative indices; that is, formulae that are symmetric in their use of weighing information and which satisfy a set of economic criteria. The Fisher and Törnqvist indices are examples of such formulae. This preference can be considered as identifying the theoretical targets to which National Statistical Institutes aspire when they go about producing price and quantity indices; however, as with most matters in the realm of index numbers, transferring theory to practice is not a simple process.

Most price indices evolved from a need to measure changing prices, often initially to compensate workers with appropriate wage increases during periods when prices were unstable. In that context there was little room for extensive debate about the validity of various axioms in the determination of an appropriate index formula. There were also issues of computational simplicity to consider in the years before computers were widely used (or used at all),[4] and issues of how well the general public can understand in broad terms how a measure of inflation is constructed. It is worth bearing in mind that the economic theory for construction of a cost-of-living index was developed after price indexes had begun to be constructed and used.

As with any statistical measure, the availability of good quality data is as important as selecting a formula and for price indices practical considerations around data play

---

[4] The UK Retail Prices Index was computerised in 1966.

a significant role. As was noted in Chapter 1, the creation of a fair measure of the level of prices requires expenditure weights and these can only be obtained from a time period prior to the base period. This makes producing a Fisher or a Törnqvist in a timely manner impractical – it also rules out producing a Paasche and a Laspeyres index too. As was noted in Chapters 5 and 12, this means that National Statistics Institutes (NSIs) use the Lowe formula to produce a measure of inflation. Some NSIs do calculate index numbers based on superlative formulae, but only on a retrospective basis, waiting several years until suitable weighting information is available [16]. More generally, the production of price indices on a timely basis requires significant compromises and measuring them against ideal, theoretical frameworks will show many differences.

The question of the best formula to use will continue to be discussed between those who favour superlative indices and those who do not; it may be that a consensus will emerge in time. Even if superlative formulae do gain (near) universal acceptance, there is, of course, the practical problem of producing them in a timely fashion. One approach would be to investigate whether they can be approximated using only base period weighting data. Another approach would be to investigate the use of scanner data; that is, large datasets of transactions from supermarkets which are made available very quickly and from which up to date weights can be obtained. These are topics that we consider in Chapter 15.

The current official position is clear; NSIs use the Lowe formula. They also update the basket and the expenditure weights on a regular basis, usually annually, and chain their index numbers. This practice reduces the difference between using a Lowe formula and a superlative formula. Some NSIs produce a superlative index version of their consumer price index on a retrospective basis to demonstrate the difference that choosing an alternative formula makes.

13.7   Chapter summary

The choice of index formula is important; different formulae produce different index numbers and therefore different values for inflation. Finding a 'best' formula is a difficult challenge and has occupied economists and statisticians for many years. Four main approaches to try to answer this question have been developed and although there is no universally accepted conclusion, many experts favour a superlative index such as the Fisher or Törnqvist. While these indices may have attractive properties they cannot currently be calculated in a timely manner and so NSIs use the Lowe formula with regular updates to the representative basket of goods and expenditure weights. A more detailed discussion of the axiomatic and economic approaches is presented in Appendix B.

References

1. Statistics Canada website (2014) Definitions, Data Sources and Methods, International Merchandise and Trade Price Index (IMTPI), Detailed information for August 2014, http://www23.statcan.gc.ca/imdb/p2SV.pl?Function=getSurvey&SDDS=2203&lang=en &db=imdb&adm=8&dis=2 (accessed 16 January 2015).

2. Fisher, I. (1922) *The Making of Index Numbers: A Study of their Varieties, Tests, and Reliability*, Houghton Mifflin, Boston, MA, http://babel.hathitrust.org/cgi /pt?id=uc2.ark:/13960/t3dz0582m;view=1up;seq=7 (accessed 16 January 2014), Fisher wrote a previous book: The Purchasing Power of Money, 1911 which also discussed index formulae; however, it is his 1922 that is devoted to the identification of preferred formulae.

3. See for example Diewert, W.E. (2013) Consumer Prices in the UK, http://www .economics.ubc.ca/files/2013/06/pdf_paper_erwin-diewert-consumer-price-statistics-UK .pdf (accessed 19 January 2015) which contains a discussion of axioms.

4. Bennet, T.L. (1920) The theory of measurement of change in the cost of living. *Journal of the Royal Statistical Society*, **83** (3), P455–462.

5. Bowley, A.A. (1919) The measurement of changes in the cost of living. *Journal of the Royal Statistical Society*, **82** (3), P343–372.

6. Konüs, A.A. (1939) The problem of the true index of the cost of living (translation from Konüs 1924). *Econometrica*, **7** (1), 10–29.

7. This is a common utility function used in introductions to mathematical economics, students wishing to find out more are referred to Varian, H.R. (2014) *Intermediate Microeconomics: A Modern Approach*, W.W. Norton & Company, New York, Or any similar economics textbook.

8. See, for example, ILO (2004) *The Consumer Price Index Manual: Theory and Practice*, International Labour Office, Individual Chapters can be Accessed from the Following Page, http://www.ilo.org/public/english/bureau/stat/guides/cpi/#manual (accessed 16 January 2014).

9. Diewert, W.E. (1976) Exact and superlative index numbers. *Journal of Econometrics*, **4** (2), 115–145. doi: 10.1016/0304-4076(76)90009-9, This is a classic paper in the theory of index numbers.

10. See, for example,Deaton, A. and Muellbauer, J. (1980) An almost ideal demand system. *The American Economic Review*, **70** (3), 312–326.

11. a See Chapter 5 Balk, B. (2008) *Price and Quantity Index Numbers*, Cambridge University Press, Cambridge, p. 175; b Also,Balk, B. (2005) Price indices for elementary aggregates. *Journal of Official Statistics*, **21** (4), 675–699.

12. Diewert, E. (2010) *On the Stochastic Approach to Index Numbers, Index Measures*, vol. **6**, Chapter 11, http://www.indexmeasures.com/ (accessed 16 January 2015). http:// bookstore.trafford.com/Products/SKU-000178116/Price-and-Productivity-Measurement-Volume-6--Index-Number-Theory.aspx.

13. Feenstra, R. and Reinsdorf, M. (2007) Should Exact Index Numbers Have Standard Errors? NBER Report, www.nber.org/papers/w10197 (accessed 16 January 2015).

14. Elliott, D., O'Neill, R., Ralph, J. and Sanderson, R. (2012) Stochastic and Sampling Approaches to the Choice of Elementary Aggregate Formula, http://www.ons .gov.uk/ons/guide-method/user-guidance/prices/cpi-and-rpi/index.html (accessed 16 January 2015).

15. For a description of the different models and index formulae, see, for example,von der Lippe, P. (2007) *Index Theory and Price Statistics*, Peter Lang, Frankfurt, p. 83.

16. The US Bureau of Labor Statistics publish have published a superlative index version of their CPI, Cage, R., Greenlees, J. and Jackman, P. (2003) Introducing the Chained Consumer Price Index, The US Bureau of Labor Statistics, http://www.bls.gov /cpi/super_paris.pdf (accessed 16 January 2015).

Exercise K

These questions are designed to encourage you to think more broadly about the topics you have covered and how they can be applied in practice. Formal solutions are not provided for these questions; however, outlines of answers can be found in the accompanying on-line content.

K.1   Which way of choosing an index number formula seems the most sensible to you?

K.2   What reasons do you have for rejecting other methods in question 1?

K.3   Given the information in this chapter what would it take before you were convinced that a given index number formula was right for measuring inflation?

K.4   What criteria would you view as most important in an index number if you were designing one from scratch?

K.5   If you had access to data regarding every economic transaction in the country how would you measure inflation?

14

Issues in index numbers

In Chapter 13, we looked at identifying the best formula to use and the attempts that have been made to come up with a definitive answer. This is a difficult topic in index numbers and has not been completely resolved. In this chapter, we introduce a number of other issues in price index construction that have proven difficult to resolve, either theoretically or practically. Like the question of the index formula, we can only give a brief introduction to each topic as each is the subject of much analysis and discussion with competing expert views. There are also practical issues to consider and what may seem like a good idea for a small-scale or theoretical situation may not be practical in an environment when an index covers a wide geographic region including several thousand prices collected each month, which must be reported to strict deadlines.

This chapter is not meant to present an exhaustive list of challenging topics of further interest in index numbers; rather, they are the issues, which experts have thought most deeply about.

14.1 Cost-of-living versus cost-of-goods

Given the importance of the measurement of inflation, it might be thought that it must surely be a well-defined, unambiguous concept; however, this is not quite the case. An effective, high-level definition can be made; for example: 'a fall in the value of money' as we specified in Chapter 3. However, when it comes to needing to define a measure and a measurement procedure, a much more detailed definition is required and this is not so straightforward.

An immediate source of confusion comes from the use of the expression 'cost of living'. The media frequently report and discuss the 'cost of living' and this term has a popular understanding as the cost of buying the goods and services that we all need. However, the same expression is used in a technical sense where it has a specific,

A Practical Introduction to Index Numbers, First Edition. Jeff Ralph, Rob O'Neill and Joe Winton.
© 2015 John Wiley & Sons, Ltd. Published 2015 by John Wiley & Sons, Ltd.
Companion Website: http://www.wiley.com/go/ralph/index_numbers

technical meaning. There are two key ways of looking at the Consumer Prices Index (CPI) statistic and they are generally known as the 'cost-of-goods' approach and the 'cost-of-living' approach. Here, we will summarise what each is and why the CPI does not currently fulfil the role of a cost-of-living measurement, meaning that the common use of the term 'cost-of-living' is not in step with the actual measurement that is made each month.

A cost-of-goods approach to a price index attempts to measure how the price of a fixed combination of goods has changed over a period of time. It is perhaps best visualised by thinking of a big shopping basket containing all of the significant items purchased by a given population; this was described in Chapter 8. This basket is then priced up and the change in prices of this fixed combination of goods is measured and used to measure inflation. This is what the CPI does, albeit using a system of weights and representative products in order to measure the changing cost of a combination of goods.

A cost-of-living approach to a price index is closely linked to the economic concept of utility and is used to represent a fixed standard of living. An economist would view individuals' purchasing behaviour in terms of utility, generally the amount of satisfaction they get from consumption of an item, or a combination of items. There are further assumptions about the behaviour of individuals, which we might make; however, the key concept here is that when we measure the cost-of-living, we are more concerned with the minimum cost of obtaining a given level of utility rather than the cost of a fixed basket of goods. As utility is not directly measurable, this raises several issues for the compilers of price indices. It also means that it is difficult to see the traditional CPI as a true cost of living index as it may be possible for the cost-of-goods index to increase by more than the cost-of-living index as people choose to change their behaviour as relative prices of goods change.

There have been several attempts to measure a cost-of-living index; perhaps the seminal contribution has been made by Diewert [1] who argues that indices such as Fisher, Törnqvist and Walsh approximate a cost-of-living index under certain conditions. Others suggest that some of the arguments made in such an approach are not realistic and, therefore, that alternative ways of measuring a cost-of-living index are preferable [2]. We do not intend to enter into this debate here, other than to note that there is still a discussion of the theoretical validity of the proposed cost-of-living indices. Most National Statistics Institutes acknowledge a cost-of-goods approach as their target though they do take guidance from the cost-of-living approach where it is helpful. The Bureau of Labor Statistics in the US does explicitly accept a cost-of-living framework as its measurement objective [3].

The difference between a cost-of-goods and a cost-of-living index might seem trivial to the outsider; however, if we are to design price indices, which are appropriate for their use, we need to understand what it is we are aiming to measure and what we are able to measure. A cost-of-living index might possess attractive attributes; however, the complexities involved in providing an accurate measure of such a statistic are not to be overlooked and are likely to generate considerable debate.

14.2   Consumer behaviour and substitution

When discussing a cost-of-living index, we mentioned the idea that as relative prices change, people might change the composition of their basket. In a simplistic example, consider that you buy both pasta and rice in your weekly shopping. If in a given week, pasta is twice as expensive as rice, you will pick a combination of the two products (along with your other shopping), which maximises your utility. If prices change so that pasta is now five times as expensive as rice, you might very well choose to substitute some of your pasta consumption for rice. Indeed your utility optimising ratio of pasta dishes to rice dishes is likely to have changed. Capturing this type of substitution behaviour is one aspect that makes it difficult to measure a cost-of-living index; however, a price index that can measure this sort of effect is of great potential interest.

The pasta and rice example was very simplistic.[1] As we move towards the real-world situation where consumers have to make decisions in the context of their complex lives, it becomes more difficult to measure how individuals will react to changes in price and what the impact is on their overall level of utility or well-being. It becomes even more complex as we start to talk about large groups of consumers rather than individuals. There are further difficulties if we allow time to enter into the problem, as well. Consumer preferences are likely to change over time, as are the selection of products available, the amount of marketing and the size of packages. While consumers do substitute goods in response to price changes, measuring and separating this from other decisions is a complex and challenging task.

At the elementary aggregate level of a price index, there is an often made argument that using the Jevons (geometric average of price relatives) rather than the Carli (the arithmetic average) is more representative of consumers' substitution behaviour, a bold claim that deserves further consideration given what we have said above about the difficulty of measuring the degree to which consumers substitute. The argument is based on a very specific model of utility, the Cobb–Douglas utility function, which we met in Section 13.3, and if we assume further that people are indifferent between products within a given elementary aggregate, then the Jevons is an exact cost-of-living index; that is, it measures the change in income required to maintain a constant level of utility. While this at first might seem useful, the Cobb–Douglas function may very well not be a good model for the real-world behaviour of consumers. Research conducted into the degree of substitution [4] suggests that people often substitute more than that would be implied by a Cobb–Douglas approach and so the argument regarding the appropriateness of the Jevons for representing substitution behaviour is questionable.

The issue of consumer substitution is directly linked to the discussion of 'cost-of-goods' versus 'cost-of living' indices mentioned above. Analysis of consumer purchase data does indicate what we would expect from our intuition, that

---

[1] For example, we assumed that the price change was material enough to you that you changed your behaviour, and also that you had access to the new prices when choosing your menu for the week.

consumers do substitute in response to changes in price. Therefore, there is a need to incorporate the effects of this in a cost-of-living index, but less need to do so, if we accept the cost-of-goods index as our measurement objective.

14.3   New and disappearing goods

In practice, index numbers are calculated from a fixed basket of goods and services that are chosen to be representative of the wide range of available goods and services that we are interested in. In theory, this is relatively simple to achieve; we pick a representative basket of goods in the base period and hold that basket constant in each subsequent period. Over time, the range of available goods and services will change; new goods will become available, other goods will no longer be available and the quality of some goods will change.

For example, if we had been compiling a price index in the 1950s, then we might have included goods such as mangles and vinyl records in our representative basket to be priced in order to measure inflation. Over the life of the price index, these products have all but disappeared[2] and so would be unlikely to be included in the basket of goods and services. The case of music media highlights this issue well as the dominant media format has changed several times over the life of national price indices: progressing from vinyl records, to cassette tapes, to compact discs, to digital downloads. How to deal with new and disappearing goods in a price index is, therefore, an important issue to consider.

There are several approaches to the inclusion of new goods within a price index. One such approach, which might be naturally favoured by an economist, is to use information about the price and demand for a new product to estimate a price at which the demand for the product would have been zero. This may be relevant in the short term; however, it remains a difficult and technically challenging area of development for those compiling price indices on a regular basis. The advent of large scanner data sets may improve the understanding of demand behaviour for new products.

A second approach to incorporating new goods into a basket of goods for use in a price index is to refresh the basket periodically. This is the approach used by most National Statistics Institutes in compiling their baskets. At periodic intervals, they will consider the items in their basket of goods and services and remove those that are no longer representative and/or no longer attract a significant proportion of expenditure.

Although the regular refreshing of the basket is a practical and established way of accommodating new products, there remain several issues with such an approach. Firstly, as the basket refresh essentially represents the start of a new price index for a new basket of goods and services, in order to create a long-term (continuous) index, there is a need to chain together the new and old index as we saw in Chapter 7. The chaining process is useful in that it permits changes in the basket and expenditure weights, but it is not without issues or critics. The chaining process will almost always yield a different set of index numbers to a direct (unchained) index depending on the

---

[2] Though there is still a lively niche market in vinyl records.

index formula used – the difference being known as 'chain-drift'. The more often the index is chained, the greater the potential for drift; however, without regularly updating the basket, the index would become unrepresentative, and so a balance is required. Most National Statistics Institutes chain every year, or every 2 years.

A further issue with new goods and services being included in an updated basket is that often the price movements of goods at the very start and the end of their lives are excluded from the price index, despite these being the periods in which their price movements are most extreme. However, it might also be noted that for most products, the expenditure weight of products is likely to be low at the beginning and the end of their lives due to the low volumes sold. It is possible to think of examples where this may not be the case; for example, around the release of new eagerly anticipated products such as new mobile phones.

The inclusion of new and disappearing goods in a price index is an important issue and highlights the fact that there are issues for which there may be better solutions in future from the study of large, detailed sets of consumer transaction data.

14.4   Quality change

When capturing prices of goods, a price collector will try to match and price an identical item each month; as a result, any change to the specifications of the item means that it is no longer the same item. The same item means one of the same quality. There are some products for which the changes in quality of the good for sale are rapid and represent a significant issue for those attempting to measure changes in prices. Examples of such products include mobile phones, computers and cameras. New models with new features are released regularly and older models disappear. In this case, the changing quality of the product represents an important challenge for those compiling price statistics.

In practice, maintaining a fixed basket, even over short periods, can be problematic; in the UK CPI, approximately 8% of all locally collected prices each month are replacement items – that is, items collected to replace items that are no longer available.[3] So what are the options when a product is no longer available?

Omitting the product is not always a good idea; with less information to use in the calculation, the resulting index number could become less accurate and less reliable. If we omit the product from the calculation, we also have a decision to make about the periods we omit the product. Do we omit the product from the point it becomes unavailable or do we revise the index in previous periods to omit the product from all time points?

Carrying forward the price of the now unavailable product is not a good idea either. The carry forward approach is to use the last price available for that product in all subsequent periods. By carrying forward the last price available for a product, it implies that there has been no change in price. This can add artificial stability to the index and the practice is banned by Eurostat [5].

---

[3] Based on analysis of published price quotes http://www.ons.gov.uk/ons/guide-method/user-guidance /prices/cpi-and-rpi/cpi-and-rpi-item-indices-and-price-quotes/index.html

Instead, finding a replacement product is generally advised as it means there are more price quotes from which to calculate an index number. However, finding a replacement can be difficult. The price collector will aim to find a product that is as similar as possible to the one it replaces. This is not always easy; no matter how well we pick a replacement product, there will always be some difference in quality between the new and the replacement item.

Consider again the UK CPI, where a basket of items is chosen in January of each year and remains fixed for the following 12 months. For each sub-index (say chocolate biscuits 300 g), a sample of distinct products matching this description is selected in January. The price of each of these products is collected every month and used to calculate the average price change for that stratum (normally using one of the un-weighted indices discussed in Chapter 4). If in a particular month, say April, a product is no longer available, then it is replaced by an alternative product; often this product is 'not comparable' to the original. It may be different in any number of ways, it may be produced by a different company, made in a different way or aimed at a different audience but still fits the broad item description – it is still a 300 g pack of chocolate biscuits. If this is the case, then the product cannot be used in the calculation of the price index for chocolate biscuits because any observed price change may be purely the result of the difference between the original product and the newly selected alternative.

Imagine, instead, that we are trying to compile a price index for a good such as mobile phone, which has a number of measurable features that determine its price, such as camera specification, screen resolution, Internet capability and operating system. A price collector would ideally like to pick one such product and regularly price that exact product every month. However, with new models introduced frequently, at some point, the specified phone will not be available any more. Should the price collector switch to an alternative phone and apply the price changes of that to the price observed for the previous item? Women's clothing presents similar difficulties as garments change regularly with the fashion season.

Adjusting prices to account for changes in quality is an important topic for index number practitioners and the ILO CPI manual [6] discusses the issues and options at length. In this section, we will discuss a few of these options.

14.4.1   Option 1: do nothing – pure price change

Doing nothing implies that the replacement item is exactly the same as the old item in every way and that the difference in price between them is a pure price change. If a good replacement is selected, then 'pure price change' may be a reasonable assumption; however, decisions about whether a replacement is comparable or not are often very subjective.

14.4.2   Option 2: automatic linking – pure quality change

One option to avoid is to assume that the difference between the price of the original and the replacement item is purely a result of a difference in quality and, therefore, that the 'true' price has not changed. When there is a large difference in price between the replacement item and the original item it has replaced, it may be reasonable to assume that there is a large difference in quality and it may not be possible to disentangle the change in price from the change in quality. In this case, the replacement item can be linked to the old item so that there is no price change in the period in which the replacement appears. This will add stability to the index series and as such is banned by Eurostat [7].

14.4.3 Option 3: linking

If the prices of both the new and the replacement items are available in some common period, then it is possible to link the two series together. Unfortunately, due to the very nature of the problem, we usually do not know that an item will need to be replaced until it disappears. This makes collecting the prices of both items for some link period very difficult. If, however, this is possible, then linking the price relatives would work in the same way as linking index series in Chapter 7 and would be a preferable option.

14.4.4 Option 4: imputation

Consider again the example of the pack of biscuits in the UK CPI. If in a particular month, a particular pack is no longer available, then it is replaced by an alternative product; this product cannot be used directly in the calculation of the price index for chocolate biscuits because any observed price change may be purely a result of the difference between the original product and the newly selected alternative and not due to a change in price.

Because the alternative product is new to the sample, a January (base) price was not collected and the product cannot be included in the index calculation in the traditional way. Instead, a base price can be imputed using the price at which the product first appeared in the sample and the price change of the rest of the products up to that point.

If a replacement item enters the basket in April, then a base price is imputed for the new item. This is achieved by applying the growth rate between the base period and April to the price of the replacement item in April (calculated using the other items in the group). This is like saying – if the price of this new item moved in exactly the same way as the rest of the basket between January and April, then what would its price have been in January? This imputed January price can then be used to calculate price changes for the replacement item between January and the months following April.

Imputation of the base prices for new products is designed in a way that makes full use of the new product in the months following its appearance, but has no effect on the index in the month that it first appears. In other words, in the month that the new product first appears in the sample, the index will be the same whether or not the new item is included.

14.4.5 Option 5: hedonics

A more sophisticated technique for managing quality change for a complex item is to collect prices for a variety of types of the item that have different values of attributes (like different resolutions of camera on a phone) and to construct a statistical model that identifies the contribution of each attribute to the overall price of the item. This is known as hedonic modelling[4]. Once a model has been determined, it can be used to price an arbitrary phone once the values of its attributes are known (for example, screen resolution and camera resolution). Although hedonic modelling can provide an effective mechanism for adjusting for changes in complex goods, it does have the disadvantage of requiring a significant amount of effort to collect data and run a statistical model, so its use is limited.

14.5   Difficult to measure items

There are some items that are required in the basket to represent total consumption but cause complex problems for the compilers of price indices. In this section, we explore some examples of commodities that should be in an index such as the CPI but where measurement of the prices is difficult.

Perhaps the most controversial item, which is difficult to measure for the purposes of estimating inflation, is the consumption of housing services. Many people in the UK live in homes, which they own; however, while they are living in those homes, they are consuming housing services during that period and an economist might say they are 'deriving utility from living in the property'. There is then the difficult question of measuring the price of this service. For people renting their accommodation, this is much simpler – the cost of the housing service they are provided with is their rent. One approach to owner-occupied housing might, therefore, be to ask how much people who live in houses they own might pay in rent to live in a similar property. There are several problems with this approach, not least that such prices would have to be imputed from rental prices and, in general, imputed prices are not allowed in price indices. Some might suggest using a house price index, such as that produced by the Office for National Statistics, to measure the changing cost of housing; however, this measures the purchase price of the entire property, which includes a significant investment element, especially as people have tended to expect prices paid for houses to steadily increase over time. Measuring the nominal value of housing services consumed by those who live in properties they own is a very difficult task, and given the potential importance of this to a measure of inflation, it is one that is hard to ignore. The current practice in the UK was described briefly in Section 8.9.

In recent years, clothing has also caused problems for those compiling official measures of price statistics. It might appear a simple good to include – we can go to a shop and price the same item continuously; however, this does not work so well for items whose design, manufacture and appeal change almost continuously. In many ways, it might seem fair to apply a method such as hedonic regression to clothing; however, there is an important unobserved variable, which seems to affect the price of clothing, that might be thought of as its 'fashionability' – a vague attribute that captures how fashionable an item is. As a result, even if we observe the same physical item in

---

[4] See Ref. [13] in Chapter 13.

subsequent periods, it may be that they are different items in the eyes of consumers and so we are no longer pricing matching items as we would like.

Most people make use of financial services such as loans, insurance and bank accounts; however, pricing these services can also be difficult. If you hold a basic current account with a bank, it is unlikely you pay a fee directly for the services associated with the account; however, you are likely to consume them. Pricing such services can be very challenging and as financial services become more complex, they are increasingly bundled together with other products.

In this section, we have introduced just a few of the challenging issues with measuring the prices of goods and services consumed for inclusion in a price index. For some goods, it is difficult to obtain objective, consistent prices for items that are not transacted in an easy way. For these items, there remains a considerable debate regarding their measurement.

14.6   Chapter summary

This chapter has described some of the challenges that index theorists and price statisticians face in defining exactly how prices should be measured in principle and how they should be collected in practice. This is an active topic, and there is a healthy research community looking at how we can make improvements to current methods. The next chapter will look at some of these research topics.

References

1. Diewert, E. (1976) Exact and superlative index numbers. *Journal of Econometrics*, **4** (2), 115–445.
2. Breuer, C. and von der Lippe, P. (2011) Problems of Operationalising a Cost of Living Index, http://mpra.ub.uni-muenchen.de/32902/ (accessed 16 January 2015).
3. The change in the measurement objective for the Bureau of Labor Statistics from a cost of goods to a cost of living occurred in 1997. This is well described in: National Research Council, Schultze, C., and Mackie, C. (2002) *At What Price, Conceptualising and Measuring Cost-of-Living and Price Indices*, National Academic Press.
4. Elliott, D. and O'Neill, R. (2012) Estimating the Elasticity of Substitution for Alcohol Products, ONS Survey Methodology Bulletin No. 21, http://www.ons.gov.uk/ons/guide -method/method-quality/survey-methodology-bulletin/smb-71/index.html (accessed 16 January 2015).
5. Eurostat (1996) HICP European Commission Regulation (EC) No. 1749/96, Article 6, http://epp.eurostat.ec.europa.eu/portal/page/portal/hicp/documents/Tab/Tab/02-IMPLEM -EN.pdf (accessed 16 January 2015).
6. See chapter 7 of the ILO (2004) *Consumer Price Index Manual: Theory and Practice*, Geneva, International Labour Office, http://www.ilo.org/public/english/bureau/stat/guides /cpi/ (accessed 16 January 2015).
7. Eurostat (1996) HICP European Commission Regulation (EC) No. 1749/96, Article 6 http://epp.eurostat.ec.europa.eu/portal/page/portal/hicp/documents/Tab/Tab/02-IMPLEM -EN.pdf (accessed 16 January 2015).

15

Research topics in index numbers

In the earlier chapters of this book, we explained how important it is to measure inflation and we have provided an introduction to some of the techniques that are required to enable National Statistics Institutes (NSIs) to produce price index numbers in a timely fashion. Over many years, professional practice and academic study have enabled 'best-practice' standards to emerge and these guide NSIs in producing price statistics in a consistent and high-quality manner [1]. However, as we saw in Chapters 13 and 14, there are many aspects of index numbers where the practical challenges are great and more research work is needed to enable NSIs to produce better measures. This chapter looks at some of those areas and gives an overview of research activities.

15.1   The uses of scanner data

Scanner data is a term that is used to describe large, detailed data sets of economic transactions. The easiest way to think of the creation of such data might be in a supermarket: when you take items to the checkout, the items are scanned electronically and the store creates an electronic record of exactly what has been bought, when and at what price. A large collection of these electronic records of transactions constitute very useful information with potential for statistical purposes. As more and more sales are recorded electronically, such data sets will cover more and more of the transactions taking place within an economy, recording both the price and the quantity of goods bought.

However, there are several problems to bear in mind with transaction data sets; they are usually the property of private companies who may not want such information made public for commercial or other reasons; the data sets are very large and require

A Practical Introduction to Index Numbers, First Edition. Jeff Ralph, Rob O'Neill and Joe Winton.
© 2015 John Wiley & Sons, Ltd. Published 2015 by John Wiley & Sons, Ltd.
Companion Website: http://www.wiley.com/go/ralph/index_numbers

appropriate storage and processing facilities and there is not yet a clear consensus about how such data should be best used in the construction of price indices.

In this section, we describe four research topics that might be furthered by the use of scanner data and outline some recent work, which has been carried out in these areas. As the availability and the use of scanner data grow, these are just some of the areas that might benefit from the use of this emerging source of data.

15.1.1   Improvements at the lowest level of aggregation

One of the most difficult aspects of price index construction occurs at the lowest level – the elementary aggregate level; elementary aggregates were described in Sections 4.2 and 8.7. At this level of the aggregation structure, we do not have expenditure weights in most cases, and price statisticians have to use simple, unweighted formulae containing just price information. If quantity information were available, the unweighted formulae could be replaced by weighted formulae, which would produce a better measure of inflation. Scanner data have the potential to provide this quantity information and so enable greater use of weighed formulae at the lowest level. It may also enable us to create a finer division of goods and services, thereby increasing homogeneity, which might be beneficial.

15.1.2   Understanding consumer behaviour

In Chapter 14 (and in Appendix B.2), we have discussed economic measures of inflation, or cost-of-living indices, and in doing so, we have stated that we need to make assumptions about how consumers behave in order to reach conclusions and lack data with which to test and refine those assumptions. Scanner data could provide the quantity data required and this would represent an important step in being able to test and refine the economic approach to index numbers.

If National Statistics Institutes could obtain scanner data on a regular basis, they could use that data to derive up-to-date quantity data and so give us the capability to calculate index numbers using weighting information from both the current and base periods. In some countries, this type of data has become available on a regular basis and progress is being made in its use. However, processing scanner data on a regular basis requires significant computing capability and staff resource. Usually, scanner data only cover a subset of all goods and services, so weights can only be obtained for a partial set of goods within the index, which itself presents a challenge. Practical experience with similar data has shown that much work is required to process the data into a form where items match strict COICOP (Classification of Individual Consumption by Purpose) product descriptions (see Section 6.1 for a brief description of the COICOP classification structure).

Research has been carried out using scanner-type data to better understand the economic behaviour of individuals in certain markets [2, 3]. This used scanner data to research consumer behaviour in the UK alcohol market, measuring the elasticity of substitution for alcohol products in the United Kingdom. The elasticity of substitution is a measure of how readily consumers change what they purchase in response to relative price changes. Using the elasticity of substitution estimates, the authors were able to show that the assumptions behind the use of a Jevons index as a cost-of-living index were not well borne out by the evidence from the detailed data.

Other researchers have also used scanner data to investigate consumer behaviour in the Netherlands [4]; for example, noting that for some products, such as washing powder, individuals most commonly only bought these items when they were on special offer. This research shows that researchers will be able to use scanner data to get more information about how people shop and how they consume.

Further research could investigate whether different types of consumers substitute in different ways; such research would help build knowledge about how consumers behave, which will be of benefit to understand the degree to which different types of consumers respond to price changes in different ways.

15.1.3   Alternative measurement schemes

Access to scanner data presents a much larger set of price quotes, which could enable price statisticians to compile probability distributions of prices and price changes in order to assess what might be a good summary measure to use in representing the central tendency of such distributions. It might also identify characteristics of the data, which have been previously ignored; for example, if the distribution has particularly thick tails or is highly skewed. This will enable us to learn more about the nature of price changes and potentially use better statistics to account for the 'price populations' from which observations of individual prices are sampled.

Most price observations for inclusion in measures of inflation are collected via a price collector who follows a methodology put in place by the NSI. Scanner data provide the opportunity to test the efficiency and accuracy of such sampling schemes for collecting price data. It may be that collection of prices once a month for a well-specified set of items produces a price index, which is an accurate estimate of the population statistic; however, it may also not be the case for all commodities and to some extent, scanner data provide some ability for NSIs to test the effects of the sampling methodologies they have chosen to implement. Scanner data, for example, allow the collection of prices on multiple days across the month and, therefore, allow for the evaluation of how this increased frequency of collection, referred to as temporal sampling, might affect the resulting index.

If it is not possible to gain a great deal from integrating scanner data into the production process for price indices, scanner data research can be used to tell NSIs more about the index construction methods they are currently employing and how the results relate to those constructed from a more comprehensive data set. It may be that the current methods are appropriate; however, research in this area may allow for the refinement of the approach currently used.

15.1.4   Frequency of indices

Price indices are currently constructed monthly in the United Kingdom; however, with the advent of scanner data, it may be possible to compile weekly or daily price indices making inflation measures much more detailed, providing potentially more detailed information about the movement of prices akin to such indices in a financial market.

Whether there is a need for daily or weekly price indices for all goods and services is unknown, especially where prices are stable. However, there may be specific goods and services for which prices are sometimes relatively volatile where a more frequent index might be beneficial. Fuel is an example where this is already the practice.

15.2   Variations on indices

Two variants of current practice have been suggested that would provide useful additional information to those who use price statistics: regional indices and indices for different socio-economic groups. In both cases, additional effort and money would be required to produce them; the current sample in the United Kingdom is not designed for these purposes.

15.2.1   Regional indices

The economy of the United Kingdom is relatively diverse; however, indices such as the Consumer Prices Index (CPI) seek to sum up in a single statistic the inflation experience of the whole country. It is already possible to see the contributions towards inflation from different categories of consumer products; however, the advent of scanner data, or extension of the existing sampling approach, also raises the potential for indices to be compiled on geographical bases. It might be very useful for people to be able to compare the rate of change in prices for consumers in London and the South East of England with those in Wales, for example. This could be combined with regional measures of pay to determine which areas have made the greatest gains in terms of improvements in real income.

As with most topics in the realm of index numbers, there are issues to be considered in the construction of regional indices. If individuals consume items in a different area to where they purchased the item, to which area should changes in the price level be assigned?

Price statistics by region would have a number of potential uses; for example, pricing decisions for some goods and services could also be based on the regional measures by companies when contracting for goods and services.

15.2.2   Variation by socio-economic group or income quantile

In addition to compiling indices for different regions of a country, it would be of interest for price statisticians to compile price statistics for people with different consumption patterns. These could be achieved by examining the inflation experiences of people in different socio-economic groups. Consider, for example, the consumption decisions faced by an elderly pensioner whose income is mainly from the state pension, a young professional and a semi-retired company director: they are likely to have different patterns of consumption and the goods they consume are likely to be of different qualities and have different characteristics. The pensioner is likely to spend a greater proportion of their income on fuel and food than the company director who may travel more and spend more of their income on leisure activities; the young professional meanwhile may make use of more recent additions to the basket of goods and services such as dating websites and e-books than either of the other two. Of course, care would be needed in specifying the groups. A pensioner group may not be an ideal choice of group as retired people differ significantly in their incomes. A more practical grouping might be achieved by considering income – perhaps in quintiles or deciles.

A rough insight into the degree to which individual experiences of inflation differ from the official measure can be obtained from examining individual expenditure. The BBC personal inflation calculator, which has been developed with Warwick University[1], illustrates this by allowing individuals to adjust the weighting used to combine the different sub-indices from the CPI to estimate the level of inflation for their personal experiences. This is a good rough and ready measure; however, the approach uses the same national inflation figures in compiling this as the published index. In compilation of a price index for a specific group, it would be better to sample items that are representative of that group's consumption, from locations that are also likely to be used by such groups. To go back to our example, the young professional and the pensioner are likely to purchase very different items from different locations in the clothing and entertainment categories and it would be better to construct a sample of price observations for use in a price index that reflects this.

As with regional price indices, there are potential applications for indices for some groups. The Government could use such indices to contribute to consideration of an appropriate uprating of benefits of many groups reliant on the state for support. Clearly, there would be issues in making sure the use of such indices is fair and consistent; however, research into the differing inflation experiences of groups within the United Kingdom would be interesting as it would reveal the extent to which an aggregate measure such as the CPI is able to accurately reflect the inflation experiences of a diverse population.

15.3   Difficult items

We have mentioned throughout this book that there are a number of items for which price changes are difficult to measure, and here we discuss research areas related to the measurement of difficult-to-measure items.

15.3.1   Clothing

Fashion goods, in particular clothing, have caused concern for the compilers of price indices for a considerable period. Research towards measuring clothing inflation

---

[1] http://www.bbc.co.uk/news/business-22523612.

using scanner-type data has been carried out [5]; Sanderson (2014) attempted to identify clusters of clothing brands between which consumers substituted using a social network visualisation. If clusters of substitutable goods could be found, then this might suggest a scheme for more easily identifying substitute items for clothing items, which often are not available and so cannot be priced under the current rules of the collection. The results suggest that this is difficult even when using a detailed clothing scanner data set with detailed product descriptions. One possible missing piece of information about fashion goods is how fashionable they are. In order to measure the price of goods of a constant quality, it is important that those goods are easily identifiable; the problem is that even if we can identify an item, which is the same in terms of colour, cut and material, it may be that it has become less or more fashionable over time and, therefore, the overall quality of the product, in the eyes of the consumer, has fallen. The impact of 'fashionability' on item prices is difficult to disentangle and is an important area for resolution in the measurement of clothing inflation. Indeed, fashion items in general, and clothing items in particular, remain a source of research topics for those interested in index numbers.

Gaining a greater understanding of the price behaviour of clothing items will help price statisticians to prioritise future research in this area. Other research has investigated a detailed scanner data set surrounding one category of clothing items ('misses tops') [6]; the results of their investigation demonstrated the challenges of producing a quality-matched price index over time; even when superlative indices were used, the results showed implausible index numbers with falls in the price of clothing of 99%. They also gather evidence of the prevalence of sale prices in this category. Such research is important as it shows that scanner data sets with full item descriptions, quantity information and price data may still not provide all that is needed for treating price movements of some items.

15.3.2   New and disappearing goods

Current practice is for goods and services to come into and out of price indices via the annual refreshing of the basket and chaining process; this is how modern items such as e-books have made it into the current price collection. In most cases, this is an effective approach; even if the most rapid price change occurs when a product is first introduced; the expenditure weight will be small as relatively few of the goods are sold at first. However, care has to be taken with new products, which are eagerly anticipated by consumers (such as some consumer electronics products) where sales volumes can grow very quickly.

A research study has used scanner data to look again at the effect that new and disappearing goods have on price indices [7]; it concludes that the effect of ignoring such goods in a cost-of-living index is an upward bias in the measurement of inflation. This is an important result and one that shows the significance of new and disappearing goods in the development of future price indices. Further research may focus on the effect of new and disappearing items in a cost-of-goods index or may develop rules for the introduction of new items into an index in a scanner data context. For example, would it be reasonable to include every new good as it becomes available in the index or should there be some rules in order to maintain the stability of the index and prevent chain drift from becoming too much of a problem? The problem of new and disappearing goods has always made the practice of compiling indices difficult, and as more data become available, more research will be needed into how to deal with such items.

15.3.3   Hedonics

The use of hedonics in price indices is not new. The practice essentially involves carrying out a regression analysis of the prices of a selection of goods with varying characteristics in order to provide an estimate of the price impact of different characteristics. The ILO manual gives the example of the use of this methodology in the pricing of computers with price regressed on characteristics such as processor speed and brand name [8]. One reason for using this approach is that it can allow products of a constant specification to be priced over the course of the year even if the exact specification itself cannot be used.

The practice of using hedonics in price indices is still relatively new; it has been used in the United Kingdom in price measurement of digital cameras, computing equipment and mobile phones [9], while in the United States, it has also been used for apparel quality adjustment [10]. The possible application of this technique to a wider set of goods is an interesting research topic. One of the arguments against the use of hedonics in the past has been the time involved in the collection of data for inclusion in the model; however, such costs may fall with the arrival of scanner data. In addition, the increased availability of software and processing tools for running such regressions would no longer represent a barrier to the wider use of such techniques; however, care would need to be taken in the process of model specification.

15.4   Chaining

The practice of chaining indices together to form a long time series is one that is well known and von der Lippe provides a thorough consideration of the practice in both theoretical and applied terms [11]. Despite the fact that many NSIs use a chaining methodology, there are still questions, which surround the practice. At present time in the United Kingdom, the CPI is chained in December and January to account for changes in the weights used in the index and in the composition of the basket, while the older Retail Prices Index is chained once in January. Little research has been carried out to investigate the chain drift phenomenon; chain drift is the difference between a direct index and a chained index. The chaining of indices where arithmetic formulae are used will cause some drift with a positive bias against a direct index. It would be interesting to investigate the variation of drift across components of the CPI and with different index formulae; some initial analysis has been carried out in this area at the ONS [12].

15.5   Some research questions

Here, we distil the discussion of the previous sections into a number of questions, which suggest the direction of future research in index numbers in the near future. This is by no means an exhaustive list; however, we believe that it covers the main current areas of interest in the index number literature.

- Can we improve price index calculation at the elementary aggregate level?

- How accurate are the current methods for estimating consumer price inflation?

- How can we construct a reliable price index for fashion goods, in particular clothing?

- What would be the impact of producing indices that vary their focus by geography and income quantile?

- What is the best way to collect prices for a measure of inflation?

- How well are economic theories of consumption borne out by the evidence in scanner data?

- How do we ensure that the chaining method used in producing indices is fair and does not influence the measure of inflation unduly?

- How do we deal with the presence of new and disappearing goods in the compilation of a price index?

It is worth mentioning that these questions do not necessarily exist in isolation from each other; for example, as we learn more about chaining, it may inform the approach to new and disappearing goods. The advent of scanner data means that the possibilities for index number research have been expanded and as a result in a few years' time, we may be able to provide more concrete answers to some of the issues that have plagued those compiling index numbers for years.

References

1. See for example ILO (2004) *Consumer Price Index Manual: Theory and Practice*, International Labour Office, Geneva.
2. Elliott, D., Winton, J., and O'Neill, R. (2013) Elementary aggregate indices and lower level substitution bias. *Statistical Journal of the IAOS*, **29** (1), 11–19.
3. Elliott, D. and O'Neill, R. (2012) Estimating the Elasticity of Substitution for Alcohol Products, ONS Survey Methodology Bulletin, No 71, pp. 13–25, http://www.ons.gov.uk /ons/guide-method/method-quality/survey-methodology-bulletin/smb-71/index.html (accessed 20 January 2015).
4. Van der Grient, H. and de Haan, J. (2010) The Use of Scanner Data in the Dutch CPI. Statistics Netherlands, http://www.cbs.nl/NR/rdonlyres/14B48428-B9CC-470F-860A -85293F2C2F1E/0/2010scannerdatadutchcpi.pdf (accessed 20 January 2015).

5. Sanderson, R, (2014) Do Consumers Substitute between Clothing Brands? Estimating the Elasticity of Substitution for a Selection of Clothing Products. ONS Survey Methodology Bulletin, No 72, pp. 57–72, http://www.ons.gov.uk/ons/guide-method/method-quality/survey-methodology-bulletin/smb-72/index.html (accessed 20 January 2015).

6. Greenlees, J. and McClelland, R. (2010) Superlative and Regression-Based Consumer Price Indices for Apparel Using US Scanner Data, Bureau of Labor Statistics, http://www.iariw.org/papers/2010/8amcclelland.pdf (accessed 20 January 2015).

7. Melser, D. (2006). Accounting for the effects of new and disappearing goods using scanner data. *Review of Income and Wealth*, **52**(4), 547-568. doi:10.1111/j.1475-4991.2006.00203.x

8. International Labour Office (2004) *Consumer Price Index Manual: Theory and Practice*, Geneva, International Labour Office, http://www.ilo.org/public/english/bureau/stat/guides/cpi/#manual (accessed 20 January 2015).

9. ONS (2014) *Consumer Price Indices Technical Manual*, Office for National Statistics, http://www.ons.gov.uk/ons/guide-method/user-guidance/prices/cpi-and-rpi/cpi-technical-manual/index.html (accessed 20 January 2015).

10. Liegey, P.R.J. (1994) Apparel price indexes: effects of hedonic adjustment. *Monthly Labor Review* (H.W. Wilson – SSA), **117**, 38.

11. von der Lippe, P. (2011) *Index Theory and Price Statistics*, Peter Lang.

12. Clews, G., Dobson-McKittrick, A. and Winton, J. (2014) Comparing Class Level Chain Drift for Different Elementary Aggregate Formulae Using Locally Collected CPI Data, http://www.ons.gov.uk/ons/guide-method/user-guidance/prices/cpi-and-rpi/comparing-class-level-chain-drift-for-different-elementary-aggregate-formulae-using-locally-collected-cpi-data.pdf (accessed 20 January 2015).

Appendix A

Mathematics for index numbers

A.1 Notation

This appendix provides additional mathematical background to the index number formulae and calculations contained in this book.

A.1.1 Summation notation

Chapter 2 introduced the concept of a basket of goods and services; this is a collection of items that most people would buy at one time or another. For example, two tins of beans, a pair of jeans and a punnet of red grapes. Each item has a price, where this price is for one of the item (i.e. a 'unit' price) and a quantity.

To continue the example, let us say a single tin of beans costs £0.40, the jeans cost £10 and the punnet of red grapes cost £2. The expenditure on each item is simply the unit price times the quantity, and the total expenditure is the sum of these three item expenditures:

$$\text{Total expenditure} = £0.40 \times 2 + £10.00 \times 1 + £2.00 \times 1 = £12.80$$

Another term used to describe the expenditure is 'value' – the value of the milk is the amount of money spent on it. Similarly, the value of these goods is the total amount spent, which is £12.80. We can extend this to cover a basket of any size.

To put this into a mathematical form, we say that a general basket of items[1] can contain N items, each with its unit price p_i and quantity q_i, where $i = 1, 2, \ldots, N$. In the simple example above, there were three items, so $N = 3$, and we label the three

---

[1] For price statistics, the basket is a conceptual one that contains services such as a haircut, as well as goods.

A Practical Introduction to Index Numbers, First Edition. Jeff Ralph, Rob O'Neill and Joe Winton.
© 2015 John Wiley & Sons, Ltd. Published 2015 by John Wiley & Sons, Ltd.
Companion Website: http://www.wiley.com/go/ralph/index_numbers

unit prices and quantities as

$$p_1 = £0.40, \quad q_1 = 2$$
$$p_2 = £10.00, \quad q_2 = 1$$
$$p_3 = £2.00, \quad q_3 = 1$$

The value of, or expenditure on, item 1, we denote by $v_1 = p_1 \cdot q_1$, and we do the same for items 2 and 3. The total value V is then the sum of the individual values:

$$V = v_1 + v_2 + v_3$$

If we write this for N items, we get the following:

$$V = \sum_{i=1}^{N} v_i = p_1 q_1 + p_2 q_2 + \cdots + p_N q_N = \sum_{i=1}^{N} p_i q_i \qquad (A1)$$

where we have used a standard piece of mathematical notation ($\sum$) to denote a sum over the N values.

As we are all aware, prices are not fixed for very long; they change over time with some fluctuations in the short term, but usually with a long-term upward trend. So, when we record a price, we do so with an indication of the time period to which it refers. For example, the time period could be a particular month of a particular year, or perhaps just a particular year.

We incorporate a time period into the notation by using an additional subscript; so the unit price of ith item becomes the unit price of the ith item at time period t: p_{ti}. We can now write the total value of a collect of N items at a time period t in the following way:

$$V_t = \sum_{i=1}^{N} v_{ti} = p_{t1} q_{t1} + p_{t2} q_{t2} + \cdots + p_{tN} q_{tN} = \sum_{i=1}^{N} p_{ti} q_{ti} \qquad (A2)$$

Index numbers are about change, and what is of more interest is the change in total value between two time periods, 0 and t:

$$\frac{V_t}{V_0} = \frac{\sum_{i=1}^{N} v_{ti}}{\sum_{i=1}^{N} v_{0i}} = \frac{p_{t1} q_{t1} + p_{t2} q_{t2} + \cdots + p_{tN} q_{tN}}{p_{01} q_{01} + p_{02} q_{02} + \cdots + p_{0N} q_{0N}} = \frac{\sum_{i=1}^{N} p_{ti} q_{ti}}{\sum_{i=1}^{N} p_{0i} q_{0i}} \qquad (A3)$$

A comparison that is frequently of interest is the change of value between a month in one year and the same month in the previous year, so the time period t might represent July 2013 and time period 0 might represent July 2012.

It is not uncommon for the limits in the summation to be missing; this is just to simplify the notation and the reader should assume the summation will run from $i = 1$ to $i = N$.

The Laspeyres and Paasche price indices are written in this summation form:

$$P_L^{0,t} = P_L(0,t) = \frac{p_{t1}q_{01} + p_{t2}q_{02} + \cdots + p_{tN}q_{0N}}{p_{01}q_{01} + p_{02}q_{02} + \cdots + p_{0N}q_{0N}} = \frac{\sum_{i=1}^{N} p_{ti}q_{0i}}{\sum_{i=1}^{N} p_{0i}q_{0i}} \qquad (A4)$$

Notice a variant in the notation has been introduced:

$$P_L^{0,t} = P_L(0,t)$$

Some books and technical papers use the second form; it is just a matter of personal preference for authors; we prefer the former.

A.1.2   An alternative representation

The mathematical way of expressing the collective value of a set of items at a specified time period in Section A.1.1 is called the 'explicit summation' approach. There is another way that is sometimes used in the index number literature; it will be familiar to students who have studied maths, science and economics – that is the use of vectors. A vector can be defined as an ordered collection of elements.

The set of N prices (at a time period t) can be thought of as a vector of prices:

$$\boldsymbol{p}_t = (p_{t1}, p_{t2}, \ldots, p_{tN})$$

The vector of prices, $\boldsymbol{p}$, is the collected set of prices for the N items, where the position in the collection is important. For example, the third position is the unit price for the third item. Similarly, we can define a vector of quantities:

$$\boldsymbol{q}_t = (q_{t1}, q_{t2}, \ldots, q_{tN})$$

The value at time period t, V_t, can be written as the 'dot product' or 'scalar product' of the price and quantity vectors:

$$V_t = \boldsymbol{p}_t \cdot \boldsymbol{q}_t$$

The 'dot product' is calculated by multiplying the corresponding elements of the price vector with the quantity and adding them together:

$$V_t = \boldsymbol{p}_t \cdot \boldsymbol{q}_t = \sum_{i=1}^{N} p_{ti}q_{ti}$$

which is the same as the result (Equation A1).

Using vector notation, the Laspeyres price index can be written as

$$P_L^{0,t} = \frac{\boldsymbol{p}_t \cdot \boldsymbol{q}_0}{\boldsymbol{p}_0 \cdot \boldsymbol{q}_0}$$

Although this is a neater representation, it is the explicit notation that is most often used in the index number literature and that is what we use in this book.

A.1.3   Geometric indices

The first part of this appendix explained why summation notation is used extensively in the theory of index numbers; it is a convenient way of summarising lengthy additions.

Index theory also uses geometric combinations of terms as well as sums. Geometric combinations are very similar to arithmetic combinations except they use products in place of sums:

$$\prod_{i=1}^{N} x_i = x_1 \times x_2 \times \cdots \times x_{N-1} \times x_N$$

The geometric mean is similar to the arithmetic mean:

$$\text{geometric mean} = \left(\prod_{i=1}^{N} x_i \right)^{1/N}$$

Most of the familiar price and quantity indices are arithmetic combinations of price relatives and expenditure weights; however, as is explained in Chapter 12, there are also geometric versions as well. Consider the Laspeyres price index, firstly in arithmetic form and then in geometric form:

$$P_{L,\text{arithmetic}}^{0,t} = \sum_{i=1}^{N} \left(\frac{p_{ti}}{p_{0i}} \right) \cdot s_{0i} \quad P_{L,\text{geometric}}^{0,t} = \prod_{i=1}^{N} \left(\frac{p_{ti}}{p_{0i}} \right)^{s_{0i}}$$

These expressions are just extensions of the simple arithmetic mean and geometric mean respectively; they are extended by the inclusion of weights.

In almost all cases, when the Laspeyres price and quantity indices are referenced, it is the *arithmetic* versions that are the subject of the discussion. In this book, if we are referring to a geometric version of an index, we use the full description; for example, the geometric Laspeyres index.

A.1.4   Harmonic indices

In addition to arithmetic and geometric combinations of numbers, there is a further possibility – the harmonic combination. For a set of N numbers,

$$X = \{x_1, \ldots, x_N\}$$

The harmonic sum and the harmonic mean are given by

$$\text{Harmonic sum} = \frac{1}{\sum_{i=1}^{N} \frac{1}{x_i}}$$

$$\text{Harmonic mean} = \frac{N}{\sum_{i=1}^{N} \frac{1}{x_i}}$$

The Paasche price and quantity indices (in their weight/price relative form) take a harmonic mean form:

$$P_P^{0,t} = \frac{1}{\sum_{i=1}^{N} s_{ti} \left(\frac{p_{ti}}{p_{0i}} \right)^{-1}}$$

The Paasche price index is a weighted harmonic mean of price relatives. This is derived from the price and quantity form of the index in Section A.2.2.

A.2   Key results

There are several results from the theory of index numbers that a student should be very familiar with. An appreciation of the mathematics of these results will assist the student greatly in understanding other parts of index number theory and in answering theory-based questions. These results do appear elsewhere in the book; they are brought together here as they are worthy of study on their own.

A.2.1   The value ratio decomposition

The ratio of two values from different time periods can be decomposed in two ways:

- the product of a Laspeyres price index and a Paasche quantity index;
- the product of a Paasche price index and a Laspeyres quantity index.

The proof of both these decompositions is given here. Note the first step, which is to multiply the value ratio by the same expression for both the numerator and denominator, that is, to multiply by 1:

$$V^{0,t} = \frac{v_t}{v_0} = \frac{\sum_{i=1}^{N} p_{ti} q_{ti}}{\sum_{i=1}^{N} p_{0i} q_{0i}} = \frac{\sum_{i=1}^{N} p_{ti} q_{ti}}{\sum_{i=1}^{N} p_{0i} q_{0i}} \cdot \left[\frac{\sum_{i=1}^{N} p_{ti} q_{0i}}{\sum_{i=1}^{N} p_{ti} q_{0i}} \right]$$

Then, swap the order of terms in the numerator, leaving the denominator unchanged:

$$= \frac{\sum_{i=1}^{N} p_{ti} q_{0i}}{\sum_{i=1}^{N} p_{0i} q_{0i}} \cdot \frac{\sum_{i=1}^{N} p_{ti} q_{ti}}{\sum_{i=1}^{N} p_{ti} q_{0i}}$$

$$= P_L^{0,t} \cdot Q_P^{0,t}$$

To see the result the other way round, do the same initial operation, but multiplying by a different term:

$$
V^{0,t} = \frac{v_t}{v_0} = \frac{\sum_{i=1}^{N} p_{ti}q_{ti}}{\sum_{i=1}^{N} p_{0i}q_{0i}} = \frac{\sum_{i=1}^{N} p_{ti}q_{ti}}{\sum_{i=1}^{N} p_{0i}q_{0i}} \cdot \left[\frac{\sum_{i=1}^{N} p_{0i}q_{ti}}{\sum_{i=1}^{N} p_{0i}q_{ti}} \right]
$$

$$
= \frac{\sum_{i=1}^{N} p_{ti}q_{ti}}{\sum_{i=1}^{N} p_{0i}q_{ti}} \cdot \frac{\sum_{i=1}^{N} p_{0i}q_{ti}}{\sum_{i=1}^{N} p_{0i}q_{0i}}
$$

$$
= P_P^{0,t} \cdot Q_L^{0,t}
$$

These two results are very important, and they are used extensively when deflating expenditure to remove the effect of inflation.

A.2.2 Converting between the two forms of price and quantity indices

A second key result is deriving the 'expenditure weight/price relative' form of a price or quantity index from the 'price and quantity' form. The algebra for this is shown below for both a price and a quantity index.

Firstly, this is the conversion for the Laspeyres price index. We start by multiplying by '1' in the numerator:

$$
P_L^{0,t} = \frac{\sum_{i=1}^{N} p_{ti}q_{0i}}{\sum_{i=1}^{N} p_{0i}q_{0i}} = \frac{\sum_{i=1}^{N} p_{ti}q_{0i}\left(\frac{p_{0i}}{p_{0i}}\right)}{\sum_{i=1}^{N} p_{0i}q_{0i}}
$$

$$
P_L^{0,t} = \frac{\sum_{i=1}^{N} p_{ti}q_{0i}\left(\frac{p_{0i}}{p_{0i}}\right)}{\sum_{i=1}^{N} p_{0i}q_{0i}} = \frac{\sum_{i=1}^{N} p_{0i}q_{0i}\left(\frac{p_{ti}}{p_{0i}}\right)}{\sum_{i=1}^{N} p_{0i}q_{0i}} = \sum_{i=1}^{N} s_{0i}\left(\frac{p_{ti}}{p_{0i}}\right)
$$

where

$$
s_{0i} = \frac{v_{0i}}{\sum_{i=1}^{N} v_{0i}} = \frac{p_{0i}q_{0i}}{\sum_{i=1}^{N} p_{0i}q_{0i}}
$$

In this form, the Laspeyres price index is a base-weighted arithmetic mean of price relatives.

Now consider the Paasche price index:

$$
P_P^{0,t} = \frac{\sum_{i=1}^{N} p_{ti}q_{ti}}{\sum_{i=1}^{N} p_{0i}q_{ti}} = \frac{\sum_{i=1}^{N} p_{ti}q_{ti}}{\sum_{i=1}^{N} p_{0i}q_{ti}\left(\frac{p_{ti}}{p_{ti}}\right)}
$$

$$
= \frac{\sum_{i=1}^{N} p_{ti}q_{ti}}{\sum_{i=1}^{N} \left(\frac{p_{0i}}{p_{ti}}\right)p_{ti}q_{ti}} = \frac{\sum_{i=1}^{N} v_{ti}}{\sum_{i=1}^{N} v_{ti}\left(\frac{p_{ti}}{p_{0i}}\right)^{-1}} = \frac{1}{\sum_{i=1}^{N} s_{ti}\left(\frac{p_{ti}}{p_{0i}}\right)^{-1}}
$$

The Paasche price index is a base-weighted harmonic mean of price relatives.

Similar results apply to the Laspeyres and Paasche quantity indices; the derivation of just the Paasche version is shown:

$$Q_P^{0,t} = \frac{\sum_{i=1}^{N} p_{ti} q_{ti}}{\sum_{i=1}^{N} p_{ti} q_{0i}} = \frac{\sum_{i=1}^{N} p_{ti} q_{ti}}{\sum_{i=1}^{N} p_{ti} q_{0i} \left(\frac{q_{ti}}{q_{ti}}\right)} = \frac{\sum_{i=1}^{N} p_{ti} q_{ti}}{\sum_{i=1}^{N} p_{ti} q_{ti} \left(\frac{q_{ti}}{q_{0i}}\right)^{-1}}$$

$$= \frac{\sum_{i=1}^{N} p_{ti} q_{ti}}{\sum_{i=1}^{N} p_{ti} q_{ti} \left(\frac{q_{ti}}{q_{0i}}\right)^{-1}} = \frac{\sum_{i=1}^{N} v_{ti}}{\sum_{i=1}^{N} v_{ti} \left(\frac{q_{ti}}{q_{0i}}\right)^{-1}} = \frac{1}{\sum_{i=1}^{N} s_{ti} \left(\frac{q_{ti}}{q_{0i}}\right)^{-1}}$$

The Paasche quantity index is a current-weighted harmonic mean of quantity relatives.

A.2.3   Other examples of the price-relative/weights

The Paasche price index can be expressed as a current-weighted, harmonic mean of price relatives. However, it can also be written as a simple, weighted arithmetic mean of price relatives, though the weights are not standard expenditure shares:

$$P_P^{0,t} = \frac{\sum_{i=1}^{N} p_{ti} q_{ti}}{\sum_{i=1}^{N} p_{0i} q_{ti}} = \frac{\sum_{i=1}^{N} p_{ti} q_{ti} \left(\frac{p_{0i}}{p_{0i}}\right)}{\sum_{i=1}^{N} p_{0i} q_{ti}} = \frac{\sum_{i=1}^{N} p_{0i} q_{ti} \left(\frac{p_{ti}}{p_{0i}}\right)}{\sum_{i=1}^{N} p_{0i} q_{ti}} = \sum_{i=1}^{N} s_{0ti}' \left(\frac{p_{0i}}{p_{0i}}\right)$$

where

$$s_{0ti}' = \frac{p_{0i} q_{ti}}{\sum_{i=1}^{N} p_{0i} q_{ti}}$$

This is a composite weight, the product of the base prices and the current quantities.[2]

A.2.4   The value ratio as a product of Fisher indices

We have seen that the decomposition of the value ratio as a product of a Laspeyres price index and a Paasche quantity index and vice versa is an important and useful result. Let us now consider the Fisher index.

Recall that the Fisher price index is the square root of the product of the Laspeyres and Paasche price indices:

$$P_F^{0,t} = \sqrt{P_L^{0,t} \cdot P_P^{0,t}} = \sqrt{\frac{\sum_{i=1}^{N} p_{ti} q_{0i}}{\sum_{i=1}^{N} p_{0i} q_{0i}} \frac{\sum_{i=1}^{N} p_{ti} q_{ti}}{\sum_{i=1}^{N} p_{0i} q_{ti}}}$$

---

[2] This appears in the Royal Statistical Society Higher Certificate Module 7 exam paper on index numbers as question 3, part 3, in 2011.

In a similar way, the Fisher quantity index is the square root of the product of the Laspeyres and Paasche quantity indices:

$$Q_F^{0,t} = \sqrt{Q_L^{0,t} \cdot Q_P^{0,t}} = \sqrt{\frac{\sum_{i=1}^{N} p_{0i}q_{ti} \; \sum_{i=1}^{N} p_{ti}q_{ti}}{\sum_{i=1}^{N} p_{0i}q_{0i} \; \sum_{i=1}^{N} p_{ti}q_{0i}}}$$

Now, we construct the square root of the product of the Fisher price index and the Fisher quantity index:

$$P_F^{0,t} \cdot Q_F^{0,t} = \sqrt{\frac{\sum_{i=1}^{N} p_{ti}q_{0i} \; \sum_{i=1}^{N} p_{ti}q_{ti}}{\sum_{i=1}^{N} p_{0i}q_{0i} \; \sum_{i=1}^{N} p_{0i}q_{ti}}} \times \sqrt{\frac{\sum_{i=1}^{N} p_{0i}q_{ti} \; \sum_{i=1}^{N} p_{ti}q_{ti}}{\sum_{i=1}^{N} p_{0i}q_{0i} \; \sum_{i=1}^{N} p_{ti}q_{0i}}}$$

$$= \sqrt{\frac{\sum_{i=1}^{N} p_{ti}q_{ti}}{\sum_{i=1}^{N} p_{0i}q_{0i}}} \sqrt{\frac{\sum_{i=1}^{N} p_{ti}q_{ti}}{\sum_{i=1}^{N} p_{0i}q_{0i}}} \sqrt{\frac{\sum_{i=1}^{N} p_{ti}q_{0i}}{\sum_{i=1}^{N} p_{ti}q_{0i}}} \sqrt{\frac{\sum_{i=1}^{N} p_{0i}q_{ti}}{\sum_{i=1}^{N} p_{0i}q_{ti}}}$$

$$= \sqrt{\frac{\sum_{i=1}^{N} p_{ti}q_{ti}}{\sum_{i=1}^{N} p_{0i}q_{0i}}} \sqrt{\frac{\sum_{i=1}^{N} p_{ti}q_{ti}}{\sum_{i=1}^{N} p_{0i}q_{0i}}} = \frac{\sum_{i=1}^{N} p_{ti}q_{ti}}{\sum_{i=1}^{N} p_{0i}q_{0i}} = \frac{v_t}{v_0} = V_{0t}$$

This shows that the square root of the product of the Fisher price index and the Fisher quantity index is the value ratio.

A.3   Index Formula Styles

There are a variety of styles in which index number formulae appear in the literature and we have used several of them in this book. We could have chosen one style to be consistent throughout; however, we thought it would help the student to see several styles. A student who looks at other index number books or research papers would then be better prepared.

A general index is often represented by a capital I; it could stand for a price index, or a quantity index and an index of another kind. We use this representation in Chapters 2, 4 and 7. A price index will almost always be represented by a capital P and a quantity or volume index by a capital Q.

Another variant arises around the use of subscripts and superscripts. For a price index with price reference period 0 and current time period t, it is most common to put these time periods as superscripts and a label identifying the type of price index as a subscript, so a Laspeyres price index is usually written as $P_L^{0,t}$, though it is sometimes written as $P_L(0t)$. In a formula which includes the two time periods and a reference to an item in a basket labelled with an i, there are three main options:

$$p_i^0 \quad or \quad p_{0i} \quad or \quad p_{i0}$$

We used the first in Chapters 2 and 4, but changed to the second for chapters 12, 13 and Appendix A. In our view, the second representation is easier on the eye when many formulae are being discussed.

As a final point, the choice of style doesn't really matter, as long as what is meant is clear.

Appendix B

Choice of index formula

B.1 The axiomatic approach to index numbers

B.1.1 An introduction to the axiomatic approach

In Section 13.2, the axiomatic or test approach to index numbers was introduced. This approach seeks to identify a number of mathematical properties, which an index number formula should fulfil if it is to be effective in measuring a change in the general level of prices. The overall aim is to identify a small set of index formulae that satisfy a large number of these properties, or ideally just one formula. In this appendix, we describe some of these properties and consider how some of the more common indices perform against these 'tests'. In some cases a property is considered to be 'essential', in which case it is called 'axiomatic'; in other cases, a property is 'desirable, but not essential', and is called a 'test'. Note that not all commentators use this distinction and refer to all properties as 'tests'.

The axiomatic approach is very definitely located within the area of inflation measurement as many of the proposed properties have been developed specifically within this framework.[1]

There is no definitive source of the properties that should be used for the evaluation of index formulae; Diewert [1] lists 20 tests, many of which we discuss in this section. The test approach was championed by Fisher [2] in his monumental work where he attempted to identify, evaluate and classify a large number of index formulae. Some experts argue that only a subset of the properties presented in Diewert are relevant; others argue they should be extended. Here, we take a neutral view seeking only to explain the most common tests.

In this appendix, we adopt notation that is common in the literature on the axiomatic approach; it is convenient to use vector notation. We denote a price index between

---

[1] The use of properties might be applicable to other areas in which index numbers are commonly used; however, their interpretation may change and their relevance to the problem in hand might be altered.

A Practical Introduction to Index Numbers, First Edition. Jeff Ralph, Rob O'Neill and Joe Winton.
© 2015 John Wiley & Sons, Ltd. Published 2015 by John Wiley & Sons, Ltd.
Companion Website: http://www.wiley.com/go/ralph/index_numbers

periods 0 and 1 as I^{01} and note that there are four potential inputs to this index[2]: the price vectors in each period, $\mathbf{P}^0$ and $\mathbf{P}^1$, and the corresponding quantity vectors, $\mathbf{Q}^0$ and $\mathbf{Q}^1$. We use the notation $\mathbf{P}^1 \gg \mathbf{P}^0$ to denote that every element of the left-hand side vector is greater than the corresponding entry in the right-hand side vector, so $\mathbf{P}^1 \gg \mathbf{P}^0$ denotes that every price has increased, similarly $\mathbf{P}^1 \equiv \mathbf{P}^0$ is used to indicate that the two vectors are identical.

B.1.2   Some axioms

We begin by looking at some properties that have proven relatively uncontroversial and have been considered as 'axioms'.

B.1.2.1   Positivity: $I^{01} > 0$

This property states that it should not be possible for a price index to have a negative value. If we consider that prices and quantities have positive values, then this seems like an obvious condition.

B.1.2.2   Continuity: I^{01} is continuous in its arguments

Like the positivity property, this proposed restriction on possible price index formulae looks completely sensible.

B.1.2.3   Unity: If $\mathbf{P}^1 \equiv \mathbf{P}^0$, then $I^{01} = 1$

This property requires that if the prices in the two periods are identical, then the index must not change. Most people would see this as a quite common sense claim to make; if none of the prices have changed, how can we justify using a price index which says that the level of prices has changed? For such a property it is useful to consider whether we think what an index would look like where this was not the case. As Diewert notes, we have not restricted quantities to be the same in both periods so this might be a source of difference; however, it is difficult to think of an index in which quantities or expenditure shares are not treated symmetrically.

B.1.2.4   Monotonicity in current prices: If $\mathbf{P}^2 \gg \mathbf{P}^1 \gg \mathbf{P}^0$, then $I^{02} > I^{01}$

That is, if all prices are increasing, then the price index should also be increasing; this seems a sensible proposition. If a price index is measuring the level of prices and if every single price which we are considering has increased, then it should follow that the level of prices increases.

B.1.2.5   Monotonicity in base prices: If $\mathbf{P}^2 \gg \mathbf{P}^0$, then $I^{01} > I^{21}$

This property says that if the current prices are unambiguously further away from the base period prices than in some other period, then the increase in the index should be

---

[2] Note that we are using 0 and 1 for our two time periods here; earlier in the book we have used 0 and t. The index number literature uses both conventions.

greater for the period in which prices have moved the most. Again this seems sensible: if prices have increased more between periods 0 and 1 than they did between periods 2 and 1 we would expect the change in the index to be greater between period 0 and 1.

B.1.2.6   Mean value property: $\min_i(p_i^1/p_i^0) \leq I^{01} \leq \max_i(p_i^1/p_i^0)$, $i = 1, \ldots, N$

This property states that the value of the index should be between the maximum and minimum values of the price relatives for the goods under consideration. Again the justification for this appeals to common sense in that if an index is measuring a change in the price level, it would make little sense for the change in that level to be greater than the change in the price of any specific component of the set of goods under consideration. All of the indexes discussed in the main portion of this book fulfil this property, and we are not aware of this property being questioned in the literature.

B.1.2.7   Proportionality: If $P^1 = \lambda P^0$, then $I^{01} = \lambda \, (\lambda > 0)$

The proportionality property requires that if the proportional change is the same for every good in the basket under consideration, or, alternatively, that all of the price relatives are the same, then the index should reflect this in taking on the value of this constant proportional change. The proportional change in prices described by this axiom is unlikely to ever be achieved in an applied setting; however, this does not mean it is not useful as a property. If the proportional change in the price of all goods were the same, then the answer to what has been the change in the level of prices would seem to be to a large extent obvious. As a result, a price index that does not fulfil this property might seem to be suspect as it does not return the answer we would expect.

B.1.2.8   Inverse proportionality: If $P^2 = \lambda P^0$, then $I^{21} = \lambda^{-1} I^{01}$

This property says that if the base prices are multiplied by a constant, the value of the index using these new base prices should be the index between periods 0 and 1 multiplied by the inverse of the constant. Note that the difference between this and the original proportionality test presented above is subtle but important. This test tells us something about what we would expect from an index in a controlled environment; again it is difficult to argue that we would not expect this property to be true in the set of circumstances described.

B.1.2.9   Time reversal: $I^{01} = 1/I^{10}$

This property says that if we switched the time superscripts on the price and quantity vectors, then we would expect the index to be the inverse of the original. Fisher [2] explains this in terms of looking at differences in prices between two cities. Suppose we have two goods that are priced at £1 in location 1 and priced at £1.50 and £0.90, respectively, in location 2. If we take the average of the price changes using location 1 as the base[3], then we see an increase by 20% in price. If we took the inverse of this, we would expect a 17% decrease in price. However, if we use location 2 prices as

---

[3] $(1.5 + 0.9)/2 = 1.2$.

the base, the average of price changes is in fact[4] 11%. This therefore seems strange as starting from different locations (or time periods to move away from Fisher's example), we get different results regarding the relative prices between the two price regimes.

It should be noted that time reversal is probably the most controversial of the properties presented in this appendix; some commentators believe it is a vital characteristic of an index number, others strongly disagree, while some are undecided. The example above showed that the Carli index does not have this property. If we had taken the geometric mean of prices,[5] using the Jevons index, then we would have fulfilled the time reversal property. At times this has been made central to the arguments for choosing the Jevons rather than the Carli.

B.1.2.10   Commodity reversal: If the order of goods in the price and quantity vectors is changed, the index should remain the same

This property may seem entirely sensible again; simply saying that the order in which the goods are referenced should not influence the value of the index. This is again a property that is difficult to picture any sensible index formula violating. However, it is possible to construct a formula which does not satisfy this property.[6]

B.1.2.11   Changes in units of measurement: A change in the units in which commodities are measured does not affect the index value

This property essentially wants to preserve the index from being affected by changes in the way prices and quantities are measured. Diewert essentially assumes that if there is a linear transformation in the price and quantity vectors, it should not affect the outcome of the price index. Again this seems a sensible property for an index to have as it implies that the measurement of inflation is not dependent on the units in which we have measured the prices and quantities.

B.1.2.12   Quantity reversal test: If we swap the quantity vectors around, it should not affect the measurement of inflation by the index

This might seem a harsh property as it is easy to show that both the Laspeyres and Paasche indices fail this test; however, the Fisher index passes. It is quite straightforward to explain the reasoning behind the Laspeyres index; however, based on this test, we might reject this index out of hand. On the other hand, this test does ensure that quantities can only enter the index formula in a symmetric manner, essentially meaning that the differences in quantities in the two time periods do not cause a difference in the measurement of inflation. Hence, this is a test that might feel harsher than some of the others, but upon closer examination might seem to make sense when deciding between a large number of potential index number formulae.

---

[4] $(0.67 + 1.11)/2 = 0.89$.

[5] $\sqrt{1.5 \times 0.9} = 1/\sqrt{0.67 \times 1.11} = 1.16$.

[6] We leave identifying such formulae as an exercise for the reader.

B.1.2.13   Paasche and Laspeyres bounding: $I^{01}_{\text{Paasche}} \leq I^{01} \leq I^{01}_{\text{Laspeyres}}$

This property formalises the finding in Section B.2 that an economic measure of inflation would fall somewhere between Passche and Laspeyres indices. Adopting this test might be thought to be promoting the economic approach and the cost-of-living index, rather than a cost of goods approach is being selected; but this is not the case – it is to be treated as just another property. Note that the Fisher formula satisfies this constraint by definition as it is the geometric mean of the Paasche and the Laspeyres.

B.1.2.14   Fixed basket: If $Q^1 \equiv Q^0$ then $I^{01} = E^1/E^0$ where $E^t = (P^t)^T Q^t$ is the period t expenditure

This property says that if the quantities do not change, then the price index must account for all of the change in the expenditure index. Again this seems sensible and Paasche, Laspeyres and Fisher all satisfy this, while the elementary aggregate formulae may not. As price is the only factor that has changed in the value equation, this would appear to be a good criterion for a price index which uses quantity information.

B.1.3   Choosing an index based on the axiomatic approach

The properties discussed earlier are far from an exhaustive list; however, we hope those presented here have provided a good start for consideration of the axiomatic approach.

The ideal position using this approach would be for experts to agree on a set of most important axioms and tests, and this would lead to identifying one single 'best' formula that satisfies them all. However, this is not the case. There is disagreement on which properties are important and which are not. Also, for every formula, it is possible to identify a property that it does not satisfy.

At this point, it is instructive to consider a set of questions about the axiomatic approach:

- Which of the tests are really vital in the construction of an index number? That is, which tests, if failed, mean rejection of the formula under consideration?

- If two indexes pass the same number of tests, are they equally valid?

- In the case where indices fail different tests, is it possible to decide which of the formulae is more right or less wrong?

- If I find any of the tests less useful, can I exclude them from the list of axioms?

- What would I do if somebody had a different list of axioms to me? How would I defend my choice of tests? Would either of us be able to prove the other is wrong?

These questions suggest we should take a less idealised approach and not reject formulae if they do not possess some of the properties. Instead, we should ask which formulae 'do well' in the properties they satisfy and therefore 'point in the direction of a particular formula'. If we take this approach, then we find that the Fisher index 'does well'.

These questions are all important for those relying on the axiomatic approach. The approach led to Fisher declaring his eponymous formula as the most preferred index number formula.[7] Subsequently, we have discovered further reasons why this index number might be a good one to use; however, elements of the test approach still find their way into modern discussions of choice in index number construction [3], so it is important for price statisticians to be sure of the importance which they attach to the test approach.

B.1.4   Conclusions

The axiomatic or test approach has not led to a definitive 'best' index formula. It does work in the sense that the properties filter out a large number of possible index formulae. However, it has not been possible to gain wide agreement on an essential set of properties, and for every index formula it is possible to find a property that it does not satisfy. However, if we accept a less rigorous approach and are willing to be flexible regarding the sets of properties we consider important, and allow formulae to fail some tests, we can identify index formulae that satisfy most properties. As Fisher noted in his 1922 work, his eponymous formula satisfies many properties.

In Section B.2, we discuss on the economic approach that the Fisher formula arises as a desirable formula using very different criteria.

B.2   The economic approach to index numbers

B.2.1   The economic approach to index numbers

In this appendix, we examine the economic approach to index numbers; in doing so we assume that readers have a basic familiarity with common economic concepts such as utility functions, cost functions and substitution[8] [4]. This section requires a more advanced mathematical treatment than other part of the book. This appendix draws on arguments developed by Konüs [5], Balk [6] and Lloyd [7].

We begin with a consumer who can pick from N different products, with each amount chosen represented by x_i where $i = 1, \ldots, N$; the overall utility level for that consumer is specified by some function of the vector of quantities chosen, $\mathbf{X}$. Hence, for some given set of quantities they achieve a certain utility; that is, for some specified vector of quantities, $\mathbf{X}^*$, the utility function, here represented by $U(\cdot)$ produces some level of utility $u^* = U(\mathbf{X}^*)$. In any given period, a consumer is then faced with the problem of choosing a vector of quantities that achieves the highest possible utility given their overall budget, or alternatively minimises the cost of achieving a given level of utility.

---

[7] See Chapter 15 in Ref. [11].

[8] Students unfamiliar with such ideas might consider Varian [4] or any textbook with a substantive microeconomic element; study of consumer preferences would be highly relevant.

In deciding which quantities to pick, the consumer needs to think about the form of their utility function and the prices of the available goods. As a result, we introduce $\mathbf{P}$, the $N \times 1$ vector of prices in which the entries correspond to the prices of the goods offered by sales. This might be complicated if prices and quantities are not independent of each other. However, for our consumer in this example, we assume this is not important for the time being. If we define the total budget of our individual to equal to some amount m, then the consumer's problem can now be specified as in Equation (B1); our consumer is then trying to choose quantities which maximise their utility given their budget.[9]

$$\mathbf{X}^* = \max_{\mathbf{X}}(U(\mathbf{X})|\mathbf{P}^T\mathbf{X} \leq m) \tag{B1}$$

Having introduced a utility function, we also now introduce the cost function which is determined by the utility function. We denote this function as $C(u, \mathbf{P})$; this function has two inputs: a utility level u and the price vector as defined earlier. Using these two inputs, the cost function determines the minimum cost of achieving a given utility level given a set of prevailing prices.

The functions defined so far are useful to us in a single period of time; if we have a consumer facing a set of prices and a given budget, we can determine how that consumer should behave optimally. This set-up is also useful to us in constructing an index number. We now consider two time periods, identified by superscripts of 0 and 1, respectively, and assume that the consumer's utility function remains unchanged across time periods. In doing this, we begin to construct an economics based measure of inflation.

If we begin in period 0, the consumer faces prices $\mathbf{P}^0$ with initial budget m^0 and so selects a vector of quantities $\mathbf{X}^0$ corresponding to some utility level u^0. By definition,

$$(\mathbf{P}^0)^T\mathbf{X}^0 = m^0 = C(u^0, \mathbf{P}^0)$$

If we then allow the prices to change in the second period to $\mathbf{P}^1$, then we can ask the following questions:

- How much would we need to increase/decrease the consumer's budget by so that they still achieve their initial utility level u^0?

- What is the vector of quantities they would choose under this new set of prices $\mathbf{X}^1$, and how does this compare to $\mathbf{X}^0$?

- How has the utility level of the consumer changed as a result of the price changes in our scenario?

All of these are interesting questions; however, we focus on the first point: how would we need to adjust the consumer's budget to make up for the effect of the price changes? More formally from the functions we have defined, we want to know what is the cost of achieving u^0 under the price regime represented by $\mathbf{P}^1$, which can be

---

[9] Here, we have used matrix notation to deal with sums of products of prices and quantity to try to make the notation easier. Readers not familiar with this notation should note that $(\mathbf{P}^0)^T\mathbf{X}^0 = \sum_{i=1}^{N} p_i^0 x_i^0$, and in other cases only the superscripts on prices and quantities are changed as appropriate.

defined by the cost function $C(u^0, \mathbf{P}^1)$ which tells us the minimum cost of achieving u^0 at these prices. Note that we are not simply asking for the cost of buying the same quantities as in the first period (this would be $(\mathbf{P}^1)^T \mathbf{X}^0$) as economists allow for substitution. Importantly, we say that as the relative prices of goods change, for example, the first good is now three times as expensive as the second, while previously it was only double the price, then the optimal combination of goods to achieve a utility level will change. It is still true that $U(\mathbf{X}^0) = u^0$. However, there may be a combination of goods for which this utility level is still achieved but at a lower cost. Our optimisation problem has therefore changed to

$$\min_{\mathbf{X}}(\mathbf{P}^1 \mathbf{X} | U(\mathbf{X}) = u^0)$$

This will yield some vector of quantities that we denote as $\mathbf{X}^*$, hence $U(\mathbf{X}^*) = u^0$ and $C(u^0, \mathbf{P}^1) = (\mathbf{P}^1)^T \mathbf{X}^*$. To return to our problem, we can now define the amount that income will have to change by to maintain our consumer's standard of living, which is described in Equation (B2).

$$I^{01}_{K-L} = \frac{C(u^0, \mathbf{P}^1)}{C(u^0, \mathbf{P}^0)} = \frac{(\mathbf{P}^1)^T \mathbf{X}^*}{(\mathbf{P}^0)^T \mathbf{X}^0} \qquad (B2)$$

This is the Laspeyres–Konüs index proposed by Konüs (1924). It is given the Laspeyres name as we have used the base period utility level to determine our measure of inflation. Using a similar logic, we therefore define the Paasche–Konüs index as

$$I^{01}_{K-P} = \frac{C(u^1, \mathbf{P}^1)}{C(u^1, \mathbf{P}^0)} = \frac{(\mathbf{P}^1)^T \mathbf{X}^1}{(\mathbf{P}^0)^T \mathbf{X}^{**}}$$

where in this equation the amount in the denominator is $\mathbf{X}^{**} = \min_{\mathbf{X}}(\mathbf{P}^0 \mathbf{X} | U(\mathbf{X}) = u^1)$, the vector of quantities that achieves the period 1 utility level at the lowest cost under period 0 prices. From these equations, the analogous Konüs–Fisher index can be defined as

$$I^{01}_{K-F} = \sqrt{I^{01}_{K-L} \times I^{01}_{K-P}}$$

This therefore represents an idealised theoretical set-up; however, we have the problem discussed in Chapter 13 that the form of the utility and cost functions are not known, and that some of the assumptions underlying them have not been tested in practice. However, the framework here gives us the opportunity to consider several interesting facets of these economic indices.

Above we stated that $\mathbf{X}^* = \min_{\mathbf{X}}(\mathbf{P}^1 \mathbf{X} | U(\mathbf{X}) = u^0)$, this is the minimum cost of achieving u^0; however, we also know that $U(\mathbf{X}^0) = u^0$ so in the peculiar case where $\mathbf{X}^* = \mathbf{X}^0$ the Konüs–Laspeyres and the traditional Laspeyres indices are the same thing. This is only likely to be true where all prices increase by a common factor. In other cases, an $\mathbf{X}^*$ is chosen that differs from $\mathbf{X}^0$, which as a result of our specified minimisation it must be the case that $(\mathbf{P}^1)^T \mathbf{X}^* \leq (\mathbf{P}^1)^T \mathbf{X}^0$. That is, the rational consumer will not choose a bundle of goods $\mathbf{X}^*$ that gives less utility than $\mathbf{X}^0$ for the prices $\mathbf{P}^1$. Knowing this, we can therefore compare our Konüs–Laspeyres index with the traditional Laspeyres. The denominators are the same, and we know that the numerator in the Laspeyres index is greater or equal to the numerator in the Konüs–Laspeyres index; therefore, we know that the Laspeyres index represents an upper bound on the Konüs–Laspeyres index, that is

$$I^{01}_{K-L} = \frac{(\mathbf{P}^1)^T \mathbf{X}^*}{(\mathbf{P}^0)^T \mathbf{X}^0} \leq \frac{(\mathbf{P}^1)^T \mathbf{X}^0}{(\mathbf{P}^0)^T \mathbf{X}^0} = I^{01}_{\text{Laspeyres}}$$

This is an important result as it tells us that if we use the Laspeyres to compensate our consumer for rising prices, then we are likely to be over-compensating them for that change, essentially giving them more money than is needed to maintain their utility, which means their overall utility will be increased by the Laspeyres.

Looking at this from the Konüs–Paasche perspective in order to get the denominator of that equation, we are looking for $\mathbf{X}^{**} = \min_{\mathbf{X}}(\mathbf{P}^0 \mathbf{X} | U(\mathbf{X}) = u^1)$, again as the utility function does not change between periods then $\mathbf{X}^1$ is still one of the vectors which satisfies the condition on the utility being the same as that experienced from the chosen quantities in period 1. As a result, $\mathbf{X}^{**}$ must be the same as $\mathbf{X}^1$ or something which costs less at period 0 prices. If $\mathbf{X}^{**} = \mathbf{X}^1$, then the Konüs–Paasche index is identical to the original Paasche index. By definition, we have $(\mathbf{P}^0)^T \mathbf{X}^{**} \leq (\mathbf{P}^0)^T \mathbf{X}^1$; that is, the denominator of the Konüs–Paasche index is no greater than that of the Paasche index. As a result, we can see that the following must be true:

$$I^{01}_{K-P} = \frac{(\mathbf{P}^1)^T \mathbf{X}^1}{(\mathbf{P}^0)^T \mathbf{X}^{**}} \geq \frac{(\mathbf{P}^1)^T \mathbf{X}^1}{(\mathbf{P}^0)^T \mathbf{X}^1} = I^{01}_{\text{Paasche}}$$

Hence, if we use the Paasche index number formula to estimate inflation, we will not be changing our consumer's income by enough to compensate for the effect of the change in prices. This is almost opposite of the case we experienced with the Laspeyres type indices where the traditional form was over-compensating consumer's form. These types of result are interesting and should be borne in mind when considering the impact of the design of index numbers.

B.2.2   A result on expenditure indices

There are other interesting results that can be found by taking an economic approach. In this section, we describe one such simple result provided by Varian [4] regarding comparisons of price indices and expenditure or value indices.

We begin by defining a consumer's expenditure index as the ratio of total amounts spent under two price regimes as below[10]:

$$E^{01} = \frac{(\mathbf{P}^1)^T \mathbf{X}^1}{(\mathbf{P}^0)^T \mathbf{X}^0}$$

As in Varian [4] we then use the theory of revealed preferences in order to determine what we can tell about a consumer's overall change in cost of living from two price

---

[10] Note that the superscript 'T' means transpose and not a time period.

indices. Revealed preference tells us that if a consumer selects a bundle of goods, or in our discussion, some vector $\mathbf{X}$, then they have demonstrated that they prefer that combination of goods to all others available at that time. As a result, we are saying that if a consumer could have afforded a particular combination of goods but did not choose that combination, then they preferred the combination they did pick. If we put this into the notation above, we are saying that if $(\mathbf{P}^1)^T\mathbf{X}^1 \geq (\mathbf{P}^1)^T\mathbf{X}^0$ then the consumer could have afforded to purchase $\mathbf{X}^0$ under the price regime $\mathbf{P}^1$ but chose not to. As a result, we must conclude that the consumer is worse off if they consume $\mathbf{X}^0$ rather than $\mathbf{X}^1$, or the pattern of period 1 consumption is revealed preferred to the period 0 consumption.

As a result of knowing that a consumer is better off in period 1 if $(\mathbf{P}^1)^T\mathbf{X}^1 \geq (\mathbf{P}^1)^T\mathbf{X}^0$, then we can note that this has implications for the Laspeyres price index. As $(\mathbf{P}^1)^T\mathbf{X}^0$ is the numerator of the Laspeyres index, it follows that

$$E^{01} = \frac{(\mathbf{P}^1)^T\mathbf{X}^1}{(\mathbf{P}^0)^T\mathbf{X}^0} \geq \frac{(\mathbf{P}^1)^T\mathbf{X}^0}{(\mathbf{P}^0)^T\mathbf{X}^0} = I^{0,1}_{\text{Laspyeres}}$$

Hence, if the expenditure index is greater than the Laspeyres price index, then we can say that people's utility has increased. This is consistent with what we concluded above that the Laspeyres is an upper bound on a cost-of-living index. However, given changes in consumer behaviour, this provides a simpler check on the effect of changes on consumers' utility as we do not need to specify a utility function.

Alternatively, if $(\mathbf{P}^0)^T\mathbf{X}^0 \geq (\mathbf{P}^0)^T\mathbf{X}^1$ and $\mathbf{X}^1$ was a possible bundle of goods at time period 0, then $\mathbf{X}^0$ has been revealed as preferred to $\mathbf{X}^1$, with the implication that the expenditure index is now lower than the Paasche index as follows:

$$E^{01} = \frac{(\mathbf{P}^1)^T\mathbf{X}^1}{(\mathbf{P}^0)^T\mathbf{X}^0} \leq \frac{(\mathbf{P}^1)^T\mathbf{X}^1}{(\mathbf{P}^0)^T\mathbf{X}^1} = I^{0,1}_{\text{Paasche}}$$

As a result, we can say that if the change in expenditure is less than the change in the Paasche index given that $\mathbf{X}^0$ and $\mathbf{X}^1$ are possible bundles of goods at time period 0, then we can be confident that people are worse off than before. Again this is a practical result which is in line with the result discussed earlier that the Paasche is a lower bound for a measure of economic inflation.

The results in this section above can be used to make it clear that we do know some things about economic measures of inflation. We know that if expenditure change is greater than the Laspeyres index, people should be better off and if it is less than the Paasche then they will be worse off. The problem is that we do not have a more precise answer about the region between these indices and the cost-of-living index. There are arguments which go further than those discussed in this book; for example, Diewert [8] shows that a Fisher index (among others) approximates a cost-of-living index based on any well-behaved utility function to a second order. This is important as in some instances, such as that of the US Labor Statistics Bureau, a retrospective Fisher index is reported as a cost-of-living index. A more precise statement is that it is an approximation to such a measure based on the best arguments available to economists in constructing a cost-of-living index.

B.2.3   Example 1: Cobb-Douglas and the Jevons index

In this section, we demonstrate a special result of the economic approach to index numbers which has been often cited as evidence that the Jevons elementary aggregate formula is able to represent the substitution behaviour of consumers. It involves us making a specific set of assumptions about the form of the utility function; specifically, assuming that it is a Cobb-Douglas function. This is a utility function that is often introduced to economists early in their training. However, after showing the result that for a specific version of this function the Jevons is the exact economic measure of inflation, we pause to consider whether this result can then be carried over to the applied practice of index numbers.

We begin by specifying the form of the Cobb-Douglas utility function; we present the three good version of the function, effectively setting $N = 3$ in the above discussion. However, the arguments made here can be expanded to a setting with more goods in a straightforward manner. In the case where there are three goods available, the Cobb-Douglas utility function is specified as

$$U(\mathbf{X}) = x_1^\alpha x_2^\beta x_3^\chi$$

where $\alpha, \beta, \chi \in (0, 1)$ are parameters describing the tastes of the consumer and without loss of generality, $\alpha + \beta + \chi = 1$. We also note that the consumer's budget in this case is assumed to be equal to m and so if we denote the price of the ith good by p_i, then

$$m = p_1 x_1 + p_2 x_2 + p_3 x_3$$

Using this framework, it can be shown that the optimal quantities of the three goods to be selected by the consumer are as follows:

$$x_1 = \frac{\alpha m}{p_1}, \quad x_2 = \frac{\beta m}{p_2}, \quad x_3 = \frac{\chi m}{p_3}$$

Note that as we assume that the taste parameters will be constant over time in our utility function, the ratio of individual goods consumed relative to other goods will vary directly as relative prices change. For example, $x_1/x_2 = (\alpha/\beta)(p_2/p_1)$; this demonstrates that what we know as the elasticity of substitution of goods is 1 in the case of the Cobb-Douglas utility function as the ratio in which relative consumption changes is directly proportional to the change in prices. This is an important implicit assumption of the Cobb-Douglas utility function and one which we will return to below.

From the above discussion of economic measures of index numbers, we are really concerned with identifying the cost function corresponding to our utility function. If we set utility equal to some value in an initial period 0, u^0, where the total budget is set to be m^0, then we can see from the above results that

$$u^0 = (x_1^0)^\alpha (x_2^0)^\beta (x_3^0)^\chi = \left(\frac{\alpha m^0}{p_1^0}\right)^\alpha \left(\frac{\beta m^0}{p_2^0}\right)^\beta \left(\frac{\chi m^0}{p_3^0}\right)^\chi \tag{B3}$$

where superscripts on prices and quantities are also used to indicate base period values. By rearranging Equation (B3), we can determine that

$$m^0 = u^0 \left(\frac{p_1^0}{\alpha} \right)^\alpha \left(\frac{p_2^0}{\beta} \right)^\beta \left(\frac{p_3^0}{\chi} \right)^\chi \qquad \text{(B4)}$$

From Equation (B4), we see that the cost here changes directly with the level of utility, that is, we have a unit utility cost of $(p_1^0/\alpha)^\alpha (p_2^0/\beta)^\beta (p_3^0/\chi)^\chi$ and so our cost function in this case is

$$C(u, \mathbf{P}^0) = u \left(\frac{p_1^0}{\alpha} \right)^\alpha \left(\frac{p_2^0}{\beta} \right)^\beta \left(\frac{p_3^0}{\chi} \right)^\chi$$

If we assume that taste parameters cannot change and we allow prices to change from $\mathbf{P}^0$ to $\mathbf{P}^1$, then we see that the cost function is then defined as

$$C(u, \mathbf{P}^1) = u \left(\frac{p_1^1}{\alpha} \right)^\alpha \left(\frac{p_2^1}{\beta} \right)^\beta \left(\frac{p_3^1}{\chi} \right)^\chi$$

Taking the ratio of these two cost functions, for any utility level, then the economic inflation measure we defined above will be

$$I^{01} = \frac{C(u, \mathbf{p}^1)}{C(u, \mathbf{p}^0)} = \left(\frac{p_1^1}{p_1^0} \right)^\alpha \left(\frac{p_2^1}{p_2^0} \right)^\beta \left(\frac{p_3^1}{p_3^0} \right)^\chi$$

If we then assume that $\alpha = \beta = \chi = (1/3)$, then

$$I^{01} = \left(\prod_{i-1}^{3} \frac{p_1^1}{p_1^0} \right)^{1/3}$$

which in the N good version of the Cobb-Douglas function becomes

$$I^{01} = \left(\prod_{i-1}^{N} \frac{p_1^1}{p_1^0} \right)^{1/N}$$

This is the geometric mean of price relatives, an index which we have previously defined as the Jevons index.

We have shown therefore that if we assume a consumer's utility function is well defined by a Cobb-Douglas utility function and that the taste parameters in such a setting are equal across goods, then the Jevons is the precise answer to the question of by what ratio should income change to maintain the utility level of our consumer. This result has often been used to make the claim that the Jevons elementary aggregate represents consumers' substitution behaviour, measuring how the ratio of goods consumed changes as relative prices change. However, it is worth noting that if we had used an alternative utility function, we would also have been able to claim the

Carli as the solution to our economic index problem.[11] The model we have used has produced a neat result, but we should be careful transferring this result into practice without interrogating the model further. If we select one particular formula for use, for example the Jevons, the choice may be for reasons other than consumer substitution. It does not follow that by using Jevons that we accept the simple model of consumer substitution that is consistent with its functional form.

In obtaining the above result we have made several assumptions, some explicit such as those about taste parameters and the form of the utility function and some implicit in the model such as the assumption about the way in which consumers substitute in reaction to changes in relative prices. We need to make judgements about the validity of these assumptions in order to apply the results to index numbers. As in other disciplines, a simplified theoretical model may not always be appropriate in an applied setting, so practitioners should be careful when borrowing such theoretical results in the construction of index numbers.

B.2.4   Example 2: CES and the Lloyd-Moulton index

In this section, we present a second utility function and show that this results in a different solution to the problem of defining an economic index. This first approach was introduced by Lloyd [7], rediscovered in Moulton [9] and considered in more detail in Balk [6]. We demonstrate how the results of the economic definition of inflation change as we alter the assumptions above so that we can further underline that the results of our investigation of the same central question change as we build up a theoretical model around it.

We use the constant elasticity of substitution (CES) utility function in this section, in the N good case this is defined as

$$U(\mathbf{X}) = \left(\sum_{i=1}^{N} b_i^{1/\sigma} \left(x_i^t \right)^{(\sigma-1)/\sigma} \right)^{\sigma/(\sigma-1)}$$

where $b_i > 0 \forall i$ are the taste parameters and we assume that $\sum_{i=1}^{N} b_i = 1$. We also introduce the parameter σ which is the elasticity of substitution discussed above in the description of the Cobb-Douglas utility function. In the CES function, we assume that the elasticity of substitution is constant, as the name suggests, but no longer assume that it is equal to 1 (which we assumed for the Cobb-Douglas utility function).

It can be shown from this[12] that the cost function for the CES function is

$$C(u, \mathbf{P}^t) = u \left(\sum_{i=1}^{N} b_i \left(p_i^t \right)^{1-\sigma} \right)^{1/(1-\sigma)}$$

---

[11] This would have required a set of Leontief preferences, in which we assume that goods are consumed in fixed ratios. This line of thinking is not pursued further here.

[12] See Balk [6] for a more thorough derivation of the results used in this discussion, here we highlight only the main results.

Hence, if we use this in our pre-defined measure of economic inflation, then we achieve the measure of inflation first introduced by Lloyd [7] and often now referred to as the Lloyd-Moulton index, following the use of this index in Moulton [9]. The Lloyd-Moulton index is therefore

$$I_{L-M}^{01} = \frac{C(u, \mathbf{P}^1)}{C(u, \mathbf{P}^0)} = \frac{\left(\sum_{i=1}^{N} b_i (p_i^1)^{1-\sigma}\right)^{1/(1-\sigma)}}{\left(\sum_{i=1}^{N} b_i (p_i^0)^{1-\sigma}\right)^{1/(1-\sigma)}} = \left(\frac{\sum_{i=1}^{N} b_i (p_i^1)^{1-\sigma}}{\sum_{i=1}^{N} b_i (p_i^0)^{1-\sigma}}\right)^{1/(1-\sigma)} \tag{B5}$$

Note that as in the case of the Cobb-Douglas index, the Lloyd-Moulton index will be the same regardless of the utility level at which we choose to evaluate it as we have again determined a unit utility cost under two different sets of prices, which forms the economic measure of inflation. The above requires us to determine the taste parameters; however, there is an alternative to this which uses optimal expenditure shares. The optimal expenditure share for the ith good in period t can be denoted as

$$s_i^t = \frac{b_i (p_i^t)^{1-\sigma}}{\sum_{j=1}^{N} b_j (p_j^t)^{1-\sigma}} \tag{B6}$$

This can sometimes be observed, for example using scanner data and does not require the estimation of the taste parameters. Then it is possible to write the Lloyd-Moulton index in a form which also does not require knowledge of the taste parameters. Noting that by rearranging Equation (B6) we can see that for each $i = 1, \dots, N$

$$\sum_{j=1}^{N} b_j (p_j^t)^{1-\sigma} = \frac{b_i (p_i^t)^{1-\sigma}}{s_i^t}$$

If we then substitute this into Equation (B3), then we can see that the Lloyd-Moulton index can be derived as follows:[13]

$$I_{L-M}^{01} = \frac{C(u, P^1)}{C(u, P^0)} = \left(\frac{\sum_{i=1}^{N} b_i (p_i^1)^{1-\sigma}}{\sum_{j=1}^{N} b_j (p_j^0)^{1-\sigma}}\right)^{1/(1-\sigma)} = \left(\sum_{i=1}^{N} \frac{b_i (p_i^1)^{1-\sigma}}{\sum_{j=1}^{N} b_j (p_j^0)^{1-\sigma}}\right)$$

$$= \left(\sum_{i=1}^{N} \frac{b_i (p_i^1)^{1-\sigma}}{b_i (p_i^0)^{1-\sigma}/s_i^0}\right)^{1/(1-\sigma)}$$

$$= \left(\sum_{i=1}^{N} s_i^0 \left(\frac{p_i^1}{p_i^0}\right)^{1-\sigma}\right)^{1/(1-\sigma)} = \left(\sum_{i=1}^{N} s_i^1 \left(\frac{p_i^0}{p_i^1}\right)^{1-\sigma}\right)^{-1/(1-\sigma)}$$

---

[13] Again see Balk [6] for more details.

Hence, we can write the Lloyd-Moulton index as a function of expenditure weightings and price relatives. The problem persists that we still require the elasticity of substitution parameter. There are several ways to estimate this as discussed in Balk [6] and Elliott *et al.* [10]; however, we do not go further into this area other than to say that obtaining the parameter value required is itself a significant statistical task.

The purpose of this section has been to show that it is possible to move away from the setting which produced the specialised result that the Jevons index is the solution that economists are looking for. An economic measure of inflation can still be found, with a little more algebra; however, we now need expenditure weights and a reliable value for the elasticity of substitution parameter. This shows part of the problem in making the economic theory a little more involved; it makes it more difficult to obtain the statistic we are interested in.

B.2.5   Issues with the economic approach

Throughout this appendix, we have discussed some of the issues involved in developing an economic measure of inflation. In doing this, it has been necessary for us to make assumptions along the way, and this can be considered one of the main issues with the economic approach. However, without making assumptions it is difficult to enter into the sort of formal model building we have considered in this appendix. As with all such mathematical models, we may find that the model only represents reality in a loose sense; however, that does not mean that the conclusions from the theories themselves are without use. From the discussions in the first two sections of this appendix, we can conclude that a cost-of-living measure will fall somewhere between the two oldest pillars of the index numbers literature in the Paasche and Laspeyres indices; this in itself is useful.

In addition, in the past, discussions about the Lloyd-Moulton index have ended with a note that expenditure shares and the elasticity of substitution are impossible to obtain from normal data collection methods. This is no longer the case; with the advent of scanner data and other more detailed data sets, it is possible to envisage a future in which indexes such as the Lloyd-Moulton index discussed above are feasible. What then remains is a discussion of whether we are happy to accept the assumptions of the model in describing the behaviour of consumers, a conversation which requires perhaps more thorough testing of the assumptions behind such theories than is currently available.

At its worst one might accuse arguments rooted in economic theory as being exercises in mathematics, which bear little relation to the applied environment they attempt to describe. In this appendix, we have attempted to present results that have some practical use. With the continued development of the theoretical framework underpinning the development of the economic approach to index numbers, it is important that the users of indices proposed in such research understand the scope and limitations of the reasoning for selecting one index over another. As a result, when considering the theory of economic measures of inflation, a solid grounding in the economic theory attached to this area is of critical importance for furthering discussion and putting the findings into appropriate practice.

References

1. Diewert, W.E. (2011) Chapter 3: The Axiomatic Approach to Bilateral Index Number Theory, http://www.economics.ubc.ca/files/2014/02/pdf_course_erwin-diewert-ECON580Ch3.pdf (accessed 20 January 2015).

2. Fisher, I. (1922) The Making of Index Numbers: A Study of Their Varieties, Tests, and Reliability, Houghton Mifflin, Boston, MA.

3. Diewert, W.E. (2012) Consumer Price Statistics in the UK, http://www.ons.gov.uk/ons/guide-method/user-guidance/prices/cpi-and-rpi/erwin-diewert-report-on-consumer-price-statistics-in-the-uk.pdf (accessed 20 January 2015).

4. Varian, H.R. (1990) Intermediate Microeconomics: A Modern Approach, Norton, New York, London.

5. Konüs, A.A. (1939) The problem of the true index of the cost of living (translation from Konüs (1924)). *Econometrica*, **7** (1), 10–29.

6. Balk, B.M. (1999) On curing the CPI's substitution and new goods bias. Paper Presented at the 5th Meeting of the Ottawa Group.

7. Lloyd, P.J. (1975) Substitution effects and biases in non-true price indices. *American Economic Review*, **65** (3), 301–313.

8. Diewert, W.E. (1976) Exact and superlative index numbers. *Journal of Econometrics*, **4** (2), 115–145. doi: 10.1016/0304-4076(76)90009-9

9. Moulton, B.R. (1996) Constant Elasticity Cost-of-living Index in Share Relative Form, Bureau of Labor Statistcs, Washington, DC.

10. Elliott, D., Winton, J., and O'Neill, R. (2013) Elementary aggregate indices and lower level substitution bias. *Statistical Journal of the IAOS*, **29** (1), 11–19.

Appendix C

Glossary of terms and formulas

C.1 Commonly used terms

Aggregate	A group of consumer products
Aggregation	The process by which price or quantity indices for smaller groups of consumer products are combined into larger groups
Axiomatic approach	The approach to index numbers that examines mathematical properties of proposed index formulae and seeks to identify preferred formulae on the basis of satisfying many tests or axioms, also known as the test approach
Base period	The time period to which the current time period is compared; for a price index, it is the '0' in 'p_0'
Basket	A set of goods and services and their quantities. It can be understood as a generalised description of a 'shopping basket', which can contain services as well as physical goods
Basket index	An index of the form: $P_{\text{basket}} = \frac{\sum_i p_{ti} q_i}{\sum_i p_{0i} q_i}$ It measures the change in the value of a basket of goods and services where the quantities are fixed between two time periods
Chain drift	The difference between a direct and chained index; usually expressed as a ratio
Chain index	An index number time series constructed by joining together shorter time series to create a long-run continuous series. It often refers to a multi-year series created by joining multiple, 1-year series
Characteristics	Attributes of a product that identify it and usually affect its price

A Practical Introduction to Index Numbers, First Edition. Jeff Ralph, Rob O'Neill and Joe Winton.
© 2015 John Wiley & Sons, Ltd. Published 2015 by John Wiley & Sons, Ltd.
Companion Website: http://www.wiley.com/go/ralph/index_numbers

Carli index	A simple price (quantity) index which is the average of price (quantity) relatives. The Carli price index has the following form: $P_{\text{Carli}} = \frac{1}{n} \sum_{i=1}^{n} \frac{p_{ti}}{p_{0i}}$
Circularity test	An important test from the axiomatic approach. It states that the index number produced from a comparison of a current and a base time period should be equal to the product of index number from a base period to an intermediate time period multiplied by an index from the intermediate time period to the current period
Commodity	Generic term for a good or service
Consumption goods and services	Goods and services purchased and consumed by consumers and not used to create other goods and services. Sometime called 'final consumption goods and services'
Cost-of-goods index	A measure that calculates the ratio of the price of a basket of goods and services between two time periods
Cost-of-living index	A measure that calculated the ratio of the minimum expenditure of a consumer to maintain their utility in response to price changes. The customer is assumed to be 'utility-maximising' and whose tastes are unchanged between the two time periods
Coverage	The set of goods and services that is actually included in a basket of goods and services. This is in contrast to the scope, which is the set of goods and services that index compilers would like to have in the basket
Current or comparison period	The time period that is being compared to the base period
Current prices	The prices that apply at the current time period
Deflation	The removal of the effect of price change on a time series of values
Dutot index	A simple price (quantity) index, which is the ratio of the average of prices (quantities) for two time periods. The Dutot price index has the following form: $P_{\text{Dutot}} = \frac{\frac{1}{n}\sum_{i=1}^{n} p_{ti}}{\frac{1}{n}\sum_{i=1}^{n} p_{0i}}$
Economic approach	The economic approach to index numbers treats price and quantity as dependent variables. The prices of goods and services are treated as being given, and the quantities are solutions to minimisation problems where consumers reduce the quantity of a good or service if its relative price rises
Elementary aggregate	The basic building block of a price index. It is the combination of price information for highly comparable items and is at such a detailed level that relative expenditure is not available
Expenditure weights	The proportion of consumer expenditure on an aggregate of goods and services; usually expressed as parts per thousand

Fisher index	The geometric mean of the Laspeyres and Paasche indices; it is a superlative index. The price index is usually written as: $P_{\text{Fisher}} = \sqrt{P_{\text{Laspeyres}} \cdot P_{\text{Paasche}}}$
Indexation	The periodic adjustment of a money amount such as a tax threshold or benefit by a price index
Index reference period	The time period where the value of the index is set to equal 100
Item	A good or a service in the basket
Jevons index	A simple price (quantity) index, which is the geometric mean of the price (quantity) relatives. The Jevons price index has the following form: $P_{\text{Jevons}} = \left(\prod_{i=1}^{n} \frac{p_{ti}}{p_{0i}} \right)^{1/n}$
Laspeyres index	A basket index. For the price index version in the price relatives and weights form, it takes the sum of price relatives weighted by the expenditure share from the base period. The Laspeyres price index has the form: $P_{\text{Laspeyres}} = \sum_{i=1}^{n} \left(\frac{p_{ti}}{p_{0i}} \right) . s_{0i} = \frac{\sum_{i=1}^{n} p_{ti}q_{0i}}{\sum_{i=1}^{n} p_{0i}q_{0i}}$
Linking	The process of joining two index series, which have one or more overlap periods. Linking is used where the component series have different aspects to their construction, which would otherwise result in a discontinuity
Lowe index	A basket index; the quantities can be taken from a different time period to the prices; it is widely used where the quantity information is taken from a time period prior to the base period. It takes the form: $P_{\text{Lowe}} = \frac{\sum_{i=1}^{n} p_{ti}q_i}{\sum_{i=1}^{n} p_{0i}q_i}$
Owner-occupied housing	Housing owner by the occupiers; the costs are often measured by estimating the cost of the equivalent rent
Paasche index	A basket index. For the price index version in the price relatives and weights form, it takes the sum of price relatives weighted by the expenditure share from the current period. The Paasche price index has the form: $P_{\text{Paasche}} = \sum_{i=1}^{n} \left(\frac{p_{ti}}{p_{0i}} \right) . s_{ti} = \frac{\sum_{i=1}^{n} p_{ti}q_{ti}}{\sum_{i=1}^{n} p_{0i}q_{ti}}$
Price reference period	The time period '0' in the term p_0 in index formulae
Price index	A mathematical formula to estimate the change in the general level of prices between two time periods. It is a means of converting money from one time period to another
Price relative	The ratio of the unit price of an item at the current period to the price reference period; the price relative for the ith item us written as $R_{0ti} = \frac{p_{ti}}{p_{0i}}$
Product	Generic term for a good or service

Quality adjustment	An adjustment to the change in the price of a commodity where characteristics have changed; it removes the element of price change resulting from the change in characteristics
Quantity relative	The ratio of the quantity of an item at the current period to the quantity reference period; the quantity relative for the ith item us written as $\widetilde{R}_{0ti} = \frac{q_{ti}}{q_{0i}}$
Rebasing	A generic process that updates one or more of the index reference period, the price reference period and the weight reference period. The exact type of rebasing has to be taken from the context
Reference population	The set of households or businesses whose purchases are of interest in the construction of an index
Representative product	A product whose change in price is taken to represent the price change of many closely related product
Scope	The set of goods and services, which is the target of the index. In practice, the measurement of price or quantity change for some of them may not be possible. See also, coverage
Stochastic approach	The approach to index numbers in which price change is modelled as an underlying 'true' price change together with an error term. It treats an instance of price change for an item as a sample of an overall population of price changes
Substitution	Consumer behaviour in which a consumer changes from one product to another in response to a relative price change, or reduces the quantity of some items and increases quantities of other similar items
Superlative index	A subset of symmetric indices. A superlative index is an index that can approximate a cost-of-living index
Symmetric index	An index in which weighting information is taken from both the base and current periods with an equal contribution from both
Törnqvist index	A symmetric and superlative index. For the price index version, it is the geometric mean of price relatives, raised to the power of the arithmetic mean of the weights at the base and comparison periods $$P_{\text{Tornqvist}} = \prod_i \left(\frac{p_{ti}}{p_{0i}} \right)^{(1/2).(s_{0i}+s_{ti})}$$
Value	The product of unit price and quantity
Weight	Sets of numbers which are used to ensure that price or quantity changes are combined in a fair way. For a price index, the price relative for an item is multiplied by the proportion of household expenditure on that item
Weight reference period	The time period from which weighting information is taken. For price and quantity indices, this period is often a year in duration and occurs before the price reference period
Young index	A basket index. The price index version is of the form: $$P_{\text{Young}} = \sum_{i=1}^{n} \left(\frac{p_{ti}}{p_{0i}} \right) \cdot s_{bi}$$

C.2   Commonly used symbols

Expression	Description	Definition
Period t	Commonly refers to the current time period	
Period 0	Commonly refers to the base time period	
x^t	The value of a data point x at time t	
$I^{0,t}$	A simple relative index for period t	$I^{0,t} = 100 \times \frac{x^t}{x^0}$
	An index number for the period t with the period 0 as the base period	
$g^{s,t}$	The change (or growth) in an index series between two periods s and t $(s < t)$	$g^{s,t} = \frac{(I^{0,t} - I^{0,s})}{I^{0,s}} \times 100$
p_i^t	The price of item i in period t	
$R_i^{0,t}$	Price relative	$R_i^{0,t} = \frac{p_i^t}{p_i^0}$
	The price change between the base and current periods for item i	
q_i^t	The quantity (or volume) of item i in period t	
$\widetilde{R}_i^{0,t}$	Quantity (or volume) relative	$\widetilde{R}_i^{0,t} = \frac{q_i^t}{q_i^0}$
	The quantity change between the base and current periods for item i	
v_i^t	The value of item i in period t	$v_i^t = p_i^t \cdot q_i^t$
V^t	The total value of a basket of n goods in period t	$V^t = \sum_{i=1}^{n} p_i^t \cdot q_i^t$

C.3   Unweighted indices (price versions only)

$P^{0,t}_{\text{Carli}}$ The Carli price index

The arithmetic mean of price relatives between the base and current periods

$$P^{0,t}_{\text{Carli}} = \frac{1}{n} \sum_{i=1}^{n} R^{0,t}_i$$

$P^{0,t}_{\text{Jevons}}$ The Jevons price index

The geometric mean of price relatives between the base and current periods

$$P^{0,t}_{\text{Jevons}} = \left(\prod_{i=1}^{n} R^{0,t}_i \right)^{1/n}$$

$P^{0,t}_{\text{Dutot}}$ The Dutot price index

The ratio of the arithmetic mean of prices in the current period to the arithmetic mean of prices in the base period

$$P^{0,t}_{\text{Dutot}} = \frac{\frac{1}{n} \sum_{i=1}^{n} P^t_i}{\frac{1}{n} \sum_{i=1}^{n} P^0_i}$$

$P^{0,t}_{\text{Harmonic}}$ The Harmonic price index

The harmonic mean of price relatives between the base and current periods

$$P^{0,t}_{\text{Harmonic}} = \frac{n}{\sum_{i=1}^{n} \frac{1}{R^{0,t}_i}}$$

$P^{0,t}_{\text{CSWD}}$ The CSWD price index:

The geometric mean of the Carli and Harmonic price indices between the base and current periods

$$P^{0,t}_{\text{CSWD}} = \sqrt{P^{0,t}_{\text{Carli}} \times P^{0,t}_{\text{Harmonic}}}$$

CSWD: Carruthers, Selwood, Ward and Dalen.

C.4   Weighted indices (price versions only)

Basket index

$$P_{\text{basket}} = \frac{\sum_i p_{ti} q_i}{\sum_i p_{0i} q_i}$$

Drobish index

$$P_{\text{Drobisch}} = \frac{1}{2}.(P_{\text{Laspeyres}} + P_{\text{Paasche}})$$

Edgeworth–Marshall index

$$P_{\text{Edgeworth–Marshall}} = \frac{\sum_i p_{ti}.\left(\frac{q_{0i}+q_{ti}}{2}\right)}{\sum_i p_{0i}.\left(\frac{q_{0i}+q_{ti}}{2}\right)}$$

Fisher index

$$P_{\text{Fisher}} = \sqrt{P_{\text{Laspeyres}} \cdot P_{\text{Paasche}}}$$

Geometric Laspeyres and Paasche indices

$$P_{G,\text{Laspeyres}} = \prod_i \left(\frac{p_{ti}}{p_{0i}}\right)^{s_{0i}}$$

$$P_{G,\text{Paasche}} = \prod_i \left(\frac{p_{ti}}{p_{0i}}\right)^{s_{ti}}$$

Laspeyres index

$$P_{\text{Laspeyres}} = \sum_{i=1}^{n} \left(\frac{p_{ti}}{p_{0i}}\right).s_{0i} = \frac{\sum_{i=1}^{n} p_{ti} q_{0i}}{\sum_{i=1}^{n} p_{0i} q_{0i}}$$

Lowe index

$$P_{\text{Lowe}} = \frac{\sum_{i=1}^{n} p_{ti} q_i}{\sum_{i=1}^{n} p_{0i} q_i}$$

Paasche index

$$P_{\text{Paasche}} = \sum_{i=1}^{n} \left(\frac{p_{ti}}{p_{0i}}\right).s_{ti} = \frac{\sum_{i=1}^{n} p_{ti} q_{ti}}{\sum_{i=1}^{n} p_{0i} q_{ti}}$$

Palgrave index

$$P_{\text{Palgrave}} = \sum_i s_{ti} \left(\frac{p_{ti}}{p_{0i}}\right)$$

Törnqvist index

$$P_{\text{Tornqvist}} = \prod_i \left(\frac{p_{ti}}{p_{0i}}\right)^{(1/2).(s_{0i}+s_{ti})}$$

Walsh index

$$P_{\text{Walsh}} = \frac{\sum_i p_{ti}.\sqrt{q_{0i}.q_{ti}}}{\sum_i p_{0i}.\sqrt{q_{0i}.q_{ti}}}$$

Young index

$$P_{\text{Young}} = \sum_{i=1}^{n} \left(\frac{p_{ti}}{p_{0i}}\right) \cdot s_{bi}$$

C.4 Weighted indices (price versions only)

Appendix D

Solutions to exercises

This Appendix provides solutions to most of the exercises.

Formal solutions are not provided for any discussion questions; these questions are designed to encourage you to think more broadly about the topics you have covered and how they can be applied in practice.

Exercise A (Chapter 2)

A.1 a. For each period, we need to calculate a simple relative index with 2005 as the base period (period 0) with x^t representing the total number of employees in period t.

$$I^{0,t} = 100 \times \frac{x^t}{x^0}$$

For 2006, this would be

$$I^{2005,2006} = 100 \times \frac{x^{2006}}{x^{2005}} = 100 \times \frac{192}{192} = 100$$

and for the rest of the series:

Period		Index (2005 base)	
2007	$= x^{2007}/x^{2005} \times 100$	$= 204/192 \times 100$	$= 106.25$
2008	$= x^{2008}/x^{2005} \times 100$	$= 224/192 \times 100$	$= 116.67$
2009	$= x^{2009}/x^{2005} \times 100$	$= 220/192 \times 100$	$= 114.58$
2010	$= x^{2010}/x^{2005} \times 100$	$= 232/192 \times 100$	$= 120.83$
2011	$= x^{2011}/x^{2005} \times 100$	$= 212/192 \times 100$	$= 110.42$
2012	$= x^{2012}/x^{2005} \times 100$	$= 228/192 \times 100$	$= 118.75$
2013	$= x^{2013}/x^{2005} \times 100$	$= 220/192 \times 100$	$= 114.58$
2014	$= x^{2014}/x^{2005} \times 100$	$= 236/192 \times 100$	$= 122.92$

A Practical Introduction to Index Numbers, First Edition. Jeff Ralph, Rob O'Neill and Joe Winton.
© 2015 John Wiley & Sons, Ltd. Published 2015 by John Wiley & Sons, Ltd.
Companion Website: http://www.wiley.com/go/ralph/index_numbers

b. The percentage change in the index between periods a and b is calculated as

$$\text{percentage change} = \frac{B - A}{A} \times 100.$$

So the percentage change in the index between 2005 and 2013 is

$$g^{2005,2013} = \frac{I^{2005,2013} - I^{2005,2005}}{I^{2005,2005}} \times 100$$

$$= \frac{114.58 - 100}{100} \times 100 = 114.58 - 100 = 14.58\%$$

c. And the percentage change in the index between 2008 and 2011 is

$$g^{2008,2011} = \frac{I^{2005,2008} - I^{2005,2011}}{I^{2005,2011}} \times 100$$

$$= \frac{110.42 - 116.67}{116.67} \times 100 = \frac{-6.25}{116.67} \times 100 = -5.36\%$$

A.2 a. For each period, we need to calculate a simple relative index with 2005/2006 as the base season with x^t representing the total number of goals in season t.

$$I^{0,t} = 100 \times \frac{x^t}{x^0}$$

For 2006/2007, this would be

$$I^{2005/2006,2006/2007} = 100 \times \frac{x^{2006/2007}}{x^{2005/06}} = 100 \times \frac{1222}{1215} = 100.58$$

and for the rest of the series:

Period	Index (2005/2006 base)
2007/ 2008	$I^{2005/2006,2007/2008} = x^{2007/2008}/x^{2005/2006} \times 100 = 1217/1217 \times 100 = 100.16$
2008/ 2009	$I^{2005/2006,2008/2009} = x^{2008/2009}/x^{2005/2006} \times 100 = 1004/1217 \times 100 = 82.63$
2009/ 2010	$I^{2005/2006,2009/2010} = x^{2009/2010}/x^{2005/2006} \times 100 = 1001/1217 \times 100 = 82.39$
2010/ 2011	$I^{2005/2006,2010/2011} = x^{2010/2011}/x^{2005/06} \times 100 = 1228/1217 \times 100 = 101.07$
2011/ 2012	$I^{2005/2006,2011/2012} = x^{2011/2012}/x^{2005/2006} \times 100 = 1215/1217 \times 100 = 100.00$
2012/ 2013	$I^{2005/2006,2012/2013} = x^{2012/2013}/x^{2005/2006} \times 100 = 1212/1217 \times 100 = 99.75$

b. The percentage change in the index between 2005/2006 and 2012/2013 is

$$g^{2005/2006,2012/2013} = \frac{(I^{2005/2006,2012/2013} - I^{2005/2006,2005/2006})}{I^{2005/2006,2005/2006}} \times 100$$

$$= \frac{99.75 - 100}{100} \times 100$$

$$= 99.75 - 100 = -0.25\%$$

c. And the percentage change in the index between 2006/2007 and 2010/2011 is

$$g^{2006/2007,2010/2011} = \frac{I^{2005/2006,2010/2011} - I^{2005/2006,2006/2007}}{I^{2005/2006,2006/2007}} \times 100$$

$$= \frac{101.07 - 100.58}{100.58} \times 100$$

$$= \frac{0.49}{100.58} \times 100 = 0.49\%$$

d. And the percentage change in the index between 2009/2010 and 2012/2013 is

$$g^{2009/2010,2012/2013} = \frac{I^{2005/2006,2012/2013} - I^{2005/2006,2009/2010}}{I^{2005/2006,2009/2010}} \times 100$$

$$= \frac{99.75 - 82.39}{82.39} \times 100 = \frac{17.37}{82.39} \times 100 = 21.08\%$$

A.3   In this question, we need to use the index series to calculate the original data series.

$$x^t = \frac{I^{0,t} \times x^s}{I^{0,s}}$$

a. We are told that x^{2014} is 33 000, so

$$x^{2009} = \frac{I^{2009,2009} \times x^{2014}}{I^{2009,2014}} = \frac{100 \times 33,000}{132} = 25,000$$

b. and the rest of the series:

Year	Annual wage		
2009			**25 000**
2010	$= x^{2014}/I^{2009,2014} \times I^{2009,2010}$	$= 33\,000/132 \times 112$	$= 28\,000$
2011	$= x^{2014}/I^{2009,2014} \times I^{2009,2010}$	$= 33\,000/132 \times 112$	$= 28\,000$
2012	$= x^{2014}/I^{2009,2014} \times I^{2009,2010}$	$= 33\,000/132 \times 120$	$= 30\,000$
2013	$= x^{2014}/I^{2009,2014} \times I^{2009,2010}$	$= 33\,000/132 \times 128$	$= 32\,000$
2014			**33 000**

Exercise C (Chapter 4)

C.1 First, we need to calculate price relatives for each item in April and May with March as the base period.

$$R_i^{0,t} = \frac{p_i^t}{p_i^0}$$

For a chocolate chip cookie in April, this would be

$$R_1^{0,1} = \frac{p_i^1}{p_i^0} = \frac{1.50}{1.20} = 1.25$$

And for the remaining items:

1	Chocolate chip cookie	$R_1^{0,1} = p_1^1/p_1^0$	$= 1.50/1.20$	$= 1.25$
2	Lemon drizzle cake	$R_2^{0,1} = p_2^1/p_2^0$	$= 1.30/1.30$	$= 1.00$
3	Caramel shortbread	$R_3^{0,1} = p_3^1/p_3^0$	$= 1.20/1.00$	$= 1.20$
4	Flapjack	$R_4^{0,1} = p_4^1/p_4^0$	$= 1.20/1.10$	$= 1.09$
5	Rocky road	$R_5^{0,1} = p_5^1/p_5^0$	$= 1.95/1.90$	$= 1.03$
1	Chocolate chip cookie	$R_1^{0,2} = p_1^2/p_1^0$	$= 1.35/1.20$	$= 1.13$
2	Lemon drizzle cake	$R_2^{0,2} = p_2^2/p_2^0$	$= 1.25/1.30$	$= 0.96$
3	Caramel shortbread	$R_3^{0,2} = p_3^2/p_3^0$	$= 1.00/1.00$	$= 1.00$
4	Flapjack	$R_4^{0,2} = p_4^2/p_4^0$	$= 1.25/1.10$	$= 1.14$
5	Rocky road	$R_5^{0,2} = p_5^2/p_5^0$	$= 2.10/1.90$	$= 1.11$

Next, we need to calculate the average price in each period.

$$\frac{1}{5}\sum_{i=1}^{5} p_i^0 = \frac{1.20 + 1.30 + 1.00 + 1.10 + 1.90}{5} = 1.30$$

$$\frac{1}{5}\sum_{i=1}^{5} p_i^1 = \frac{1.50 + 1.30 + 1.20 + 1.20 + 1.95}{5} = 1.43$$

$$\frac{1}{5}\sum_{i=1}^{5} p_i^2 = \frac{1.35 + 1.25 + 1.00 + 1.25 + 2.10}{5} = 1.39$$

a. The Carli price index is an arithmetic mean of price relatives

$$P_{\text{Carli}}^{0,1} = \frac{1}{5}\sum_{i=1}^{5} R_i^{0,1} = \frac{1.25 + 1.00 + 1.20 + 1.09 + 1.03}{5} = 1.11$$

$$P_{\text{Carli}}^{0,2} = \frac{1}{5}\sum_{i=1}^{5} R_i^{0,2} = \frac{1.13 + 0.96 + 1.00 + 1.14 + 1.11}{5} = 1.07$$

b. The Jevons price index is a geometric mean of price relatives

$$P_{\text{Jevons}}^{0,1} = \left(\prod_{i=1}^{5} R_i^{0,1}\right)^{1/5} = (1.25 \times 1.00 \times 1.20 \times 1.09 \times 1.03)^{1/5} = 1.11$$

$$P_{\text{Jevons}}^{0,2} = \left(\prod_{i=1}^{5} R_i^{0,2}\right)^{1/5} = (1.13 \times 0.96 \times 1.00 \times 1.14 \times 1.11)^{1/5} = 1.06$$

c. The Dutot price index is a ratio of the arithmetic mean of prices

$$P_{\text{Dutot}}^{0,1} = \frac{\frac{1}{5}\sum_{i=1}^{5} p_i^1}{\frac{1}{5}\sum_{i=1}^{5} p_i^0} = \frac{1.43}{1.30} = 1.10$$

$$P_{\text{Dutot}}^{0,2} = \frac{\frac{1}{5}\sum_{i=1}^{5} p_i^2}{\frac{1}{5}\sum_{i=1}^{5} p_i^0} = \frac{1.39}{1.30} = 1.07$$

d. The Harmonic price index is a harmonic mean of price relatives

$$P_{\text{Harmonic}}^{0,1} = \frac{5}{\sum_{i=1}^{5} \frac{1}{R_i^{0,1}}} = \frac{5}{\left(\frac{1}{1.25} + \frac{1}{1.00} + \frac{1}{1.20} + \frac{1}{1.09} + \frac{1}{1.03}\right)} = 1.11$$

$$P_{\text{Harmonic}}^{0,2} = \frac{5}{\sum_{i=1}^{5} \frac{1}{R_i^{0,2}}} = \frac{5}{\left(\frac{1}{1.13} + \frac{1}{0.96} + \frac{1}{1.00} + \frac{1}{1.14} + \frac{1}{1.11}\right)} = 1.06$$

e. The CSWD price index is the geometric mean of Carli and Harmonic price indices

$$P_{\text{CSWD}}^{0,1} = \sqrt{P_{\text{Carli}}^{0,1} \times P_{\text{Harmonic}}^{0,1}} = \sqrt{1.11 \times 1.11} = 1.11$$

$$P_{\text{CSWD}}^{0,2} = \sqrt{P_{\text{Carli}}^{0,2} \times P_{\text{Harmonic}}^{0,2}} = \sqrt{1.07 \times 1.06} = 1.06$$

C.2 a.

$$R_{\text{Piano}}^{0,1} = \frac{p_{\text{Piano}}^1}{p_{\text{Piano}}^0} = \frac{22}{15} = 1.47$$

$$R_{\text{Guitar}}^{0,1} = \frac{p_{\text{Guitar}}^1}{p_{\text{Guitar}}^0} = \frac{16}{15} = 1.07$$

b.

$$\widetilde{R}_{\text{Piano}}^{0,1} = \frac{q_{\text{Piano}}^1}{q_{\text{Piano}}^0} = \frac{186}{268} = 0.69$$

$$\widetilde{R}_{\text{Guitar}}^{0,1} = \frac{q_{\text{Guitar}}^1}{q_{\text{Guitar}}^0} = \frac{269}{208} = 1.29$$

c. The price of piano lessons has risen much higher than the price of guitar lessons. Over the same period, the number of piano lessons has dramatically fallen, and there has been a rise in the number of guitar lessons.
Have students switched to guitar lessons because piano lessons have become more expensive?

C.3 a.

$$P_{\text{Dutot}}^{0,1} = 100 \times \frac{\frac{1}{5}\sum_{i=1}^{5} p_i^1}{\frac{1}{5}\sum_{i=1}^{5} p_i^0} = 100 \times \frac{250.00 + 7.30 + 5.50 + 5.70 + 6.50}{180.00 + 6.70 + 5.00 + 5.20 + 5.90}$$

$$= 135.6$$

b.

$$P_{\text{Jevons}}^{0,1} = 100 \times \left(\prod_{i=1}^{5} R_i^{0,1} \right)^{1/5}$$

$$= 100 \times \left(\frac{250.00}{180.00} \times \frac{7.30}{6.70} \times \frac{5.50}{5.00} \times \frac{5.70}{5.20} \times \frac{6.50}{5.90} \right)^{1/5}$$

$$= 100 \times (1.389 \times 1.090 \times 1.100 \times 1.096 \times 1.102)^{1/5} = 115.00$$

c. The price change measured by the Dutot index is far higher than that measured by the Jevons index. This is because of the large price movement in the high-priced vintage champagne. You can see from the price relatives calculated as part of the Jevons index that while the other four items all rose in price by about 10%, the price of the champagne rose by around 39%.
The large price of the champagne then dominates the averages calculated as part of the Dutot index.

Exercise D (Chapter 5)

D.1 To calculate Laspeyres and Paasche indices, we need to calculate four terms

$$\sum_i p_i^0 q_i^0 = (p_1^0 \times q_1^0) + (p_2^0 \times q_2^0) + (p_3^0 \times q_3^0)$$

$$= (35 \times 200) + (30 \times 250) + (18 \times 50)$$

$$= 15400$$

$$\sum_i p_i^1 q_i^1 = (p_1^1 \times q_1^1) + (p_2^1 \times q_2^1) + (p_3^1 \times q_3^1)$$

$$= (37.5 \times 100) + (25 \times 300) + (18 \times 45)$$

$$= 12060$$

$$\sum_i p_i^0 q_i^1 = (p_1^0 \times q_1^1) + (p_2^0 \times q_2^1) + (p_3^0 \times q_3^1)$$

$$= (35 \times 100) + (30 \times 300) + (18 \times 45)$$

$$= 13310$$

$$\sum_i p_i^1 q_i^0 = (p_1^1 \times q_1^0) + (p_2^1 \times q_2^0) + (p_3^1 \times q_3^0)$$

$$= (37.5 \times 200) + (25 \times 250) + (18 \times 50)$$

$$= 14650$$

a.

$$P_{\text{Laspeyres}}^{0,1} = 100 \times \frac{\sum_{i=1}^n p_i^1 q_i^0}{\sum_{i=1}^n p_i^0 q_i^0} = 100 \times \frac{14560}{15400} = 95.13$$

b.

$$P_{\text{Paasche}}^{0,t} = 100 \times \frac{\sum_{i=1}^n p_i^t q_i^t}{\sum_{i=1}^n p_i^0 q_i^t} = 100 \times \frac{12060}{13310} = 90.61$$

c.

$$Q_{\text{Laspeyres}}^{0,t} = 100 \times \frac{\sum_{i=1}^n p_i^0 q_i^t}{\sum_{i=1}^n p_i^0 q_i^0} = 100 \times \frac{13310}{15400} = 86.43$$

d.

$$Q_{\text{Paasche}}^{0,t} = 100 \times \frac{\sum_{i=1}^n p_i^t q_i^t}{\sum_{i=1}^n p_i^t q_i^0} = 100 \times \frac{12060}{14650} = 82.32$$

D.2

$$\sum_i p_i^0 q_i^0 = (35 \times 2) + (15 \times 13) + (7 \times 40) + (22 \times 32)$$

$$= 1249$$

$$\sum_i p_i^1 q_i^1 = (50 \times 50) + (20 \times 28) + (7 \times 18) + (18 \times 3)$$

$$= 3240$$

$$\sum_i p_i^0 q_i^1 = (35 \times 50) + (15 \times 28) + (7 \times 18) + (22 \times 3)$$

$$= 2362$$

$$\sum_i p_i^1 q_i^0 = (50 \times 2) + (20 \times 13) + (7 \times 40) + (18 \times 32)$$

$$= 1216$$

a.

$$P^{0,1}_{\text{Laspeyres}} = 100 \times \frac{\sum_{i=1}^{n} p_i^1 q_i^0}{\sum_{i=1}^{n} p_i^0 q_i^0} = 100 \times \frac{1216}{1249} = 97.36$$

b.

$$P^{0,t}_{\text{Paasche}} = 100 \times \frac{\sum_{i=1}^{n} p_i^t q_i^t}{\sum_{i=1}^{n} p_i^0 q_i^t} = 100 \times \frac{3240}{2362} = 137.17$$

c.

$$Q^{0,t}_{\text{Laspeyres}} = 100 \times \frac{\sum_{i=1}^{n} p_i^0 q_i^t}{\sum_{i=1}^{n} p_i^0 q_i^0} = 100 \times \frac{2362}{1249} = 189.11$$

d.

$$Q^{0,t}_{\text{Paasche}} = 100 \times \frac{\sum_{i=1}^{n} p_i^t q_i^t}{\sum_{i=1}^{n} p_i^t q_i^0} = 100 \times \frac{3240}{1216} = 266.45$$

e. The answers to (c) and (d) agree that the quantities of meat sold rose considerably between June and December.

The two items that experience large increases in quantity also experienced a large increase in price. This meant that the current period values for these items were very large. This causes both the Paasche price index and the Paasche quantity index to reflect the large changes in price and quantity for turkey and ham.

In this case, the Laspeyres price index is lower than the Paasche equivalent. This is an unusual result and comes from the fact that there is a positive correlation between price and quantity movements.

D.3

$$P^{0,t}_{\text{Laspeyres}} = 100 \times \sum_{i=1}^{5} R_i^{0,t} \times s_i^0$$

$$= 100 \times ((1.34 \times 0.2) + (1.22 \times 0.1) + (1.32 \times 0.35)$$

$$+ (1.01 \times 0.05) + (1.12 \times 0.3))$$

$$= 123.85$$

D.4 Let e_i^t be the number of eggs laid on farm i in period t, c_i^t be the number of chickens on farm i in period t and r_i^t be the egg rate on farm i in period t.

First, we need to calculate the egg rate r_i^t for each farm

$$r_i^t = \frac{e_i^t}{c_i^t}$$

$$r^{\text{Mar}}_{\text{Farm A}} = \frac{e^{\text{Mar}}_{\text{Farm A}}}{c^{\text{Mar}}_{\text{Farm A}}} = \frac{698}{23} = 30.35$$

	Eggs per chicken (r_i^t)	
	March	April
Farm A	30.35	25.54
Farm B	27.45	34.56
Farm C	26.45	27.60
Farm D	19.55	19.85
Farm E	28.71	28.55

a.

$$I_{\text{Laspeyres egg rate}}^{\text{Mar,Apr}} = 100 \times \frac{\sum_{i=1}^{5} c_i^{\text{Mar}} \times r_i^{\text{Apr}}}{\sum_{i=1}^{5} c_i^{\text{Mar}} \times r_i^{\text{Mar}}}$$

$$= 100 \times \frac{\begin{array}{c}(23 \times 25.54) + (31 \times 34.56) + (84 \times 27.60) \\ + (58 \times 19.85) + (21 \times 28.55)\end{array}}{\begin{array}{c}(23 \times 30.35) + (31 \times 27.45) + (84 \times 26.45) \\ + (58 \times 19.55) + (21 \times 28.71)\end{array}}$$

$$= 103.99$$

b.

$$I_{\text{Paasche egg rate}}^{\text{Mar, Apr}} = 100 \times \frac{\sum_{i=1}^{5} c_i^{\text{Apr}} \times r_i^{\text{Apr}}}{\sum_{i=1}^{5} c_i^{\text{Apr}} \times r_i^{\text{Mar}}}$$

$$= 100 \times \frac{\begin{array}{c}(28 \times 25.54) + (25 \times 34.56) + (92 \times 27.60) \\ + (61 \times 19.85) + (22 \times 28.55)\end{array}}{\begin{array}{c}(28 \times 30.35) + (25 \times 27.45) + (92 \times 26.45) \\ + (61 \times 19.55) + (22 \times 28.71)\end{array}}$$

$$= 102.81$$

c. A Laspeyres index for the number of chickens would show the effect of the change in the number of chickens in the five farms on the change in the number of eggs laid by fixing the egg rate in the base period.

d.

$$I_{\text{Laspeyres chickens}}^{\text{Mar, Apr}} = 100 \times \frac{\sum_{i=1}^{5} c_i^{\text{Apr}} \times r_i^{\text{Mar}}}{\sum_{i=1}^{5} c_i^{\text{Mar}} \times r_i^{\text{Mar}}}$$

$$= 100 \times \frac{\begin{array}{c}(28 \times 30.35) + (25 \times 27.45) + (92 \times 26.45) \\ + (61 \times 19.55) + (22 \times 28.71)\end{array}}{\begin{array}{c}(23 \times 30.35) + (31 \times 27.45) + (84 \times 26.45) \\ + (58 \times 19.55) + (21 \times 28.71)\end{array}}$$

$$= 105.19$$

Exercise E (Chapter 6)

E.1   Calculate the Laspeyres price index for each category

a.

$$P^{2013,2014}_{\text{Laspeyres; beer}} = 100 \times \sum_{i=1}^{3} R_i^{2013,2014} \times s_i^{2013}$$

$$= 100 \times ((1.25 \times 0.45) + (1.30 \times 0.20) + (1.20 \times 0.35))$$

$$= 124.25$$

$$P^{2013,2014}_{\text{Laspeyres; wine}} = 100 \times \sum_{i=1}^{5} R_i^{2013,2014} \times s_i^{2013}$$

$$= 100 \times ((1.01 \times 0.30) + (1.10 \times 0.15) + (1.05 \times 0.20)$$

$$+ (1.00 \times 0.20) + (1.12 \times 0.15))$$

$$= 104.60$$

$$P^{2013,2014}_{\text{Laspeyres; spirits}} = 100 \times \sum_{i=1}^{3} R_i^{2013,2014} \times s_i^{2013}$$

$$= 100 \times ((0.95 \times 0.60) + (0.97 \times 0.30) + (0.92 \times 0.10))$$

$$= 95.30$$

b.

$$P^{2013,2014}_{\text{Laspeyres}} = \sum_{d=1}^{3} P^{2013,2014}_{\text{Laspeyres};d} \times s_d^{2013}$$

$$= ((P^{2013,2014}_{\text{Laspeyres; beer}} \times s_{\text{beer}}^{2013}) + (P^{2013,2014}_{\text{Laspeyres; wine}} \times s_{\text{wine}}^{2013})$$

$$+ (P^{2013,2014}_{\text{Laspeyres; spirits}} \times s_{\text{spirits}}^{2013}))$$

$$= ((124.25 \times 0.50) + (104.60 \times 0.30) + (95.30 \times 0.20))$$

$$= 112.57$$

E.2   a.

$$P^{\text{Jan, May}}_{\text{Fisher; newspapers}} = \sqrt{P^{\text{Jan, May}}_{\text{Laspeyres; newspapers}} \times P^{\text{Jan,May}}_{\text{Paasch; newspapers}}}$$

$$= \sqrt{100 \times \frac{\sum_{i=1}^{3} p_i^{\text{May}} q_i^{\text{Jan}}}{\sum_{i=1}^{3} p_i^{\text{Jan}} q_i^{\text{Jan}}} \times 100 \times \frac{\sum_{i=1}^{3} p_i^{\text{May}} q_i^{\text{May}}}{\sum_{i=1}^{3} p_i^{\text{Jan}} q_i^{\text{May}}}}$$

$$= \sqrt{\begin{array}{l} 100 \times \dfrac{(1.00 \times 150) + (0.35 \times 300) + (0.65 \times 50)}{(0.90 \times 150) + (0.35 \times 300) + (0.40 \times 50)} \\[2ex] \times 100 \times \dfrac{(1.00 \times 145) + (0.35 \times 350) + (0.65 \times 26)}{(0.90 \times 145) + (0.35 \times 350) + (0.40 \times 26)} \end{array}}$$

$$= \sqrt{110.57 \times 107.97}$$

$$= 109.27$$

b.

$$P^{\text{Jan, May}}_{\text{Fisher; magazines}} = \sqrt{P^{\text{Jan, May}}_{\text{Laspeyres; magazines}} \times P^{\text{Jan,May}}_{\text{Paasch; magazines}}}$$

$$= \sqrt{100 \times \frac{\sum_{i=1}^{4} P_i^{\text{May}} q_i^{\text{Jan}}}{\sum_{i=1}^{4} P_i^{\text{Jan}} q_i^{\text{Jan}}} \times 100 \times \frac{\sum_{i=1}^{4} P_i^{\text{May}} q_i^{\text{May}}}{\sum_{i=1}^{4} P_i^{\text{Jan}} q_i^{\text{May}}}}$$

$$= \sqrt{106.66 \times 105.34}$$

$$= 106.00$$

c. to calculate the Fisher price index for all items, we have to return to the original price quotes

$$P^{\text{Jan, May}}_{\text{Fisher}} = \sqrt{P^{\text{Jan, May}}_{\text{Laspeyres}} \times P^{\text{Jan, May}}_{\text{Paasche}}}$$

$$= \sqrt{100 \times \frac{\sum_{i=1}^{7} P_i^{\text{May}} q_i^{\text{Jan}}}{\sum_{i=1}^{7} P_i^{\text{Jan}} q_i^{\text{Jan}}} \times 100 \times \frac{\sum_{i=1}^{7} P_i^{\text{May}} q_i^{\text{May}}}{\sum_{i=1}^{7} P_i^{\text{Jan}} q_i^{\text{May}}}}$$

$$= \sqrt{107.07 \times 105.62}$$

$$= 106.34$$

E.3   There are many ways to split these items. Here is one suggestion. The decisions have been made based on 'similar' sounding items and similar movements in prices.

Art
Watercolour Print, Framed Photo and Photograph Print
Crockery
Mug, Tea Pot, Spoon Set and Tea Set
Souvenirs
Key Ring, Post card, Guidebook, Coaster and Sweets.

Exercise F (Chapter 7)

F.1   2013Q1 is the link period.
   Up to the link period, the linked series is the same as the original series. From the link period onwards, the values of each series in the link period are used to calculate a link factor:

		Linked series		
2012	Q1	$= I^{2012Q1,2012Q1}$		$= 100.00$
	Q2	$= I^{2012Q1,2012Q2}$		$= 101.30$
	Q3	$= I^{2012Q1,2012Q3}$		$= 100.80$
	Q4	$= I^{2012Q1,2012Q4}$		$= 102.60$
2013	Q1	$= I^{2013Q1,2013Q1} \times I^{2012Q1,2013Q1} / I^{2013Q1,2013Q1}$	$= 100.0 \times 103.1/100.0$	$= 103.10$
	Q2	$= I^{2013Q1,2013Q2} \times I^{2012Q1,2013Q1} / I^{2013Q1,2013Q1}$	$= 101.6 \times 103.1/100.0$	$= 104.75$
	Q3	$= I^{2013Q1,2013Q3} \times I^{2012Q1,2013Q1} / I^{2013Q1,2013Q1}$	$= 100.5 \times 103.1/100.0$	$= 103.62$
	Q4	$= I^{2013Q1,2013Q4} \times I^{2012Q1,2013Q1} / I^{2013Q1,2013Q1}$	$= 102.9 \times 103.1/100.0$	$= 106.09$

F.2   In this question, we link the index in each period onto the index in the previous period:

$$I^t_{\text{Linked}} = \frac{I^{t-1,t} \times I^{t-1}_{\text{Linked}}}{100}$$

$$I^1_{\text{Linked}} = 100.0$$

$$I^2_{\text{Linked}} = \frac{I^{1,2} \times I^1_{\text{Linked}}}{I^{1,1}} = \frac{100.6 \times 100.0}{100.0} \qquad = 100.6$$

$$I^3_{\text{Linked}} = \frac{I^{2,3} \times I^2_{\text{Linked}}}{I^{2,2}} = \frac{103.0 \times 100.6}{100.0} \qquad = 103.6$$

$$I^4_{\text{Linked}} = \frac{I^{3,4} \times I^3_{\text{Linked}}}{I^{3,3}} = \frac{101.3 \times 103.6}{100.0} \qquad = 105.0$$

$$I^5_{\text{Linked}} = \frac{I^{4,5} \times I^4_{\text{Linked}}}{I^{4,4}} = \frac{103.7 \times 105.0}{100.0} \qquad = 108.8$$

$$I^6_{\text{Linked}} = \frac{I^{5,6} \times I^5_{\text{Linked}}}{I^{5,5}} = \frac{101.3 \times 108.8}{100.0} \qquad = 110.3$$

$$I^7_{\text{Linked}} = \frac{I^{6,7} \times I^6_{\text{Linked}}}{I^{6,6}} = \frac{100.7 \times 110.3}{100.0} \qquad = 111.0$$

F.3   a.

	Linked series		
2001Q1	$= I^{2001Q1,2001Q1}$		$= 100.0$
2001Q2	$= I^{2001Q1,2001Q2}$		$= 103.0$
2001Q3	$= I^{2001Q1,2001Q3}$		$= 104.0$
2001Q4	$= I^{2001Q1,2001Q4}$		$= 103.0$
2002Q1	$= I^{2002Q1,2002Q1} \times I^{2001Q1,2002Q1} / I^{2002Q1,2002Q1}$	$= 100.0 \times 108.0/100.0$	$= 108.0$
2002Q2	$= I^{2002Q1,2002Q2} \times I^{2001Q1,2002Q1} / I^{2002Q1,2002Q1}$	$= 104.0 \times 108.0/100.0$	$= 112.3$
2002Q3	$= I^{2002Q1,2002Q3} \times I^{2001Q1,2002Q1} / I^{2002Q1,2002Q1}$	$= 104.0 \times 108.0/100.0$	$= 112.3$
2002Q4	$= I^{2002Q1,2002Q4} \times I^{2001Q1,2002Q1} / I^{2002Q1,2002Q1}$	$= 104.0 \times 108.0/100.0$	$= 112.3$
2003Q1	$= I^{2002Q1,2003Q1} \times I^{2001Q1,2002Q1} / I^{2002Q1,2002Q1}$	$= 106.0 \times 108.0/100.0$	$= 114.5$

b.

	Linked series		
2001Q1	$= I^{2001Q1,2001Q1}$		$= 100.0$
2001Q2	$= I^{2001Q1,2001Q2}$		$= 103.0$
2001Q3	$= I^{2001Q1,2001Q3}$		$= 104.0$
2001Q4	$= I^{2001Q1,2001Q4}$		$= 103.0$
2002Q1	$= I_{linked}^{2002Q1}$		$= 108.0$
2002Q2	$= I_{linked}^{2002Q2}$		$= 112.3$
2002Q3	$= I_{linked}^{2002Q3}$		$= 112.3$
2002Q4	$= I_{linked}^{2002Q4}$		$= 112.3$
2003Q1	$= I^{2003Q1,2003Q1} \times I_{linked}^{2003Q1}/I^{2003Q1,2003Q1}$	$= 100.0 \times 114.5/100.0$	$= 114.5$
2003Q2	$= I^{2003Q1,2003Q2} \times I_{linked}^{2003Q1}/I^{2003Q1,2003Q1}$	$= 105.0 \times 114.5/100.0$	$= 120.2$
2003Q3	$= I^{2003Q1,2003Q3} \times I_{linked}^{2003Q1}/I^{2003Q1,2003Q1}$	$= 103.0 \times 114.5/100.0$	$= 117.9$
2003Q4	$= I^{2003Q1,2003Q4} \times I_{linked}^{2003Q1}/I^{2003Q1,2003Q1}$	$= 105.0 \times 114.5/100.0$	$= 120.2$

Exercise H (Chapter 9)

H.1 The price relatives with 2007 and Carli index with 2007 as the base year are as follows:

	Bus	Train	Taxi	Carli
2007	1.00	1.00	1.00	100.00
2008	1.50	1.10	1.09	122.90
2009	1.58	1.40	1.22	140.02
2010	1.83	1.60	1.26	156.47
2011	2.75	1.70	1.39	194.71
2012	3.75	1.80	1.48	234.28
2013	3.83	2.10	1.65	252.85
2014	4.00	2.20	1.74	264.64

H.2 The re-referenced series is estimated as follows:

	$2007 = 100$		$2012 = 100$
2007	100.00	$(100.00/234.38) \times 100 =$	42.68
2008	122.90	$(122.90/234.38) \times 100 =$	52.46
2009	140.02	$(140.02/234.38) \times 100 =$	59.77
2010	156.47	$(156.47/234.38) \times 100 =$	66.79
2011	194.71	$(194.71/234.38) \times 100 =$	83.11
2012	234.28	$(234.28/234.38) \times 100 =$	100.00
2013	252.85	$(252.85/234.38) \times 100 =$	107.93
2014	264.64	$(264.64/234.38) \times 100 =$	112.96

H.3 Using the first index number series, the percentage change between 2009 and 2013 is $((252.85/140.02) - 1) \times 100 = 81\%$ while using the second we have $((95.55/52.91) - 1) \times 100 = 81\%$. Hence, the percentage change is preserved.

H.4 The new Carli index rebased using 2010 is

	$2010 = 100$
2007	65.45
2008	78.93
2009	90.14
2010	100.00
2011	122.20
2012	144.76
2013	157.13
2014	164.54

H.5 The percentage change between 2009 and 2013 in the rebased series is $((157.13/90.14) - 1) \times 100 = 74\%$ which is not the same as the percentage change calculated in question 3.

Exercise I (Chapter 10)

I.1 a. To create a value of sales series a constant, 2007 prices, we use Equation (10.9) that expresses the value of sales at constant 2007 prices in terms of the value of sales at current price and the Paasche price index:

$$\mathrm{KP}^{t:2007} = v_t \cdot \frac{P_P^{2007,2007}}{P_P^{2007,t}} = \frac{v_t}{P_P^{2007,t}}$$

So, we need to divide the value of sales (at current prices) by the Paasche price index:

Year	Value of sales	Paasche price index	Constant 2007 prices
2012	23.54	1.432	16.44
2011	22.28	1.302	17.11
2010	21.99	1.271	17.30
2009	20.31	1.182	17.18
2008	19.76	1.101	17.95
2007	18.52	1.000	18.52

b. To calculate a Laspeyres quantity index, we use the value ratio decomposition result:

$$\frac{v_t}{v_{2007}} = P_P^{2007,t} \cdot Q_L^{2007,t} \qquad Q_L^{2007,t} = \frac{v_t}{v_{2007}} \cdot \frac{1}{P_P^{2007,t}}$$

Year	Value of sales	Paasche price index	Laspeyres quantity index
2012	23.54	1.432	0.888
2011	22.28	1.302	0.924
2010	21.99	1.271	0.934
2009	20.31	1.182	0.928
2008	19.76	1.101	0.969
2007	18.52	1.000	1.000

c. The value of sales series (at current prices) showed a consistent increase from 2007 to 2012. However, this could be the result of an increase in the volume of sales, or the increase in prices or both. The Laspeyres quantity index shows a decline in quantity of sales for most years. This means that the increase in the value of sales (at current prices) results from an increase in prices, not from an increase in quantity. This is also shown by the value of sales at constant prices.

d. To calculate the value of sales at constant, 2010 prices, we can use either formula (10.9), which uses the Paasche price index, or formula (10.10) which uses the Laspeyres quantity index:

$$KP^{t:2010} = v_t \cdot \frac{P_P^{2007,t}}{P_P^{2007,2010}} = v_{2007} \cdot \frac{Q_L^{2007,t}}{Q_L^{2007,r}}$$

The solution uses the Paasche price index approach:

Year	Value of sales	Paasche price index	Constant 2010 prices
2012	23.54	1.432	26.52
2011	22.28	1.302	22.82
2010	21.99	1.271	21.99
2009	20.31	1.182	18.89
2008	19.76	1.101	17.12
2007	18.52	1.000	14.57

I.2   To calculate the value of sales at 2008 prices given a Laspeyres quantity index, we use Equation (10.10):

$$KP^{t:2008} = v_{2008} \cdot \frac{Q_L^{2004,t}}{Q_L^{2004,8}}.$$

Year	Value of sales (CYP)	Laspeyres quantity index	Value of sales (KP, 2008)
2009	1487	0.978	1493
2008	1465	0.974	1465
2007	1452	0.968	1443
2006	1398	0.943	1354
2005	1382	0.921	1307
2004	1366	0.906	1271

Exercise J (Chapter 12)

J.1   a. Firstly, multiply the numerator by a factor that equals 1:

$$P_{\text{Basket}} = \frac{\sum_{i=1}^{N} p_{ti}q_i \cdot \left(\frac{p_{0i}}{p_{0i}}\right)}{\sum_{i=1}^{N} p_{0i}q_i}$$

Now, re-arrange the numerator into a price relative and a 'weight':

$$P_{\text{Basket}} = \frac{\sum_{i=1}^{N} \left(\frac{p_{ti}}{p_{0i}}\right) p_{0i}q_i}{\sum_{i=1}^{N} p_{0i}q_i} = \sum_{i=1}^{N} \left(\frac{p_{ti}}{p_{0i}}\right).w'_{oi} \quad \text{where } w'_{oi} = \frac{p_{0i}q_i}{\sum_{i=1}^{N} p_{0i}q_i}$$

The weight, w'_{0i}, is a composite term consisting of the ith price at time period 0 and a quantity for an arbitrary time period.

b. Firstly, multiply the Lowe index by a factor that equals 1:

$$P_{\text{Lowe}} = \frac{\sum_{i=1}^{N} p_{ti}q_{bi}}{\sum_{i=1}^{N} p_{0i}q_{bi}} = \frac{\sum_{i=1}^{N} p_{ti}q_{bi}}{\sum_{i=1}^{N} p_{0i}q_{bi}} \cdot \frac{\sum_{i=1}^{N} p_{bi}q_{ib}}{\sum_{i=1}^{N} p_{bi}q_{bi}}$$

Now, swap the first summation in the denominator with the second and then invert the second fraction and divide by it:

$$P_{\text{Lowe}} = \frac{P_L^{b,t}}{P_L^{b,0}}$$

c. The geometric Laspeyres and geometric Paasche price indices are given by

$$P_{GL}^{0,t} = \left(\prod_{i=1}^{N} \frac{p_{ti}}{p_{0i}} \right)^{w_{0i}} \qquad P_{GP}^{0,t} = \left(\prod_{i=1}^{N} \frac{p_{ti}}{p_{0i}} \right)^{w_{ti}}$$

Take the geometric mean of these two indices:

$$\sqrt{P_{GL}^{0,t} \cdot P_{GP}^{0,t}} = \sqrt{\left(\prod_{i=1}^{N} \frac{p_{ti}}{p_{0i}} \right)^{w_{0i}} \cdot \left(\prod_{i=1}^{N} \frac{p_{ti}}{p_{0i}} \right)^{w_{ti}}}$$

Simplifying, we get

$$\left(\prod_{i=1}^{N} \frac{p_{ti}}{p_{0i}} \right)^{w_{0i}/2} \cdot \left(\prod_{i=1}^{N} \frac{p_{ti}}{p_{0i}} \right)^{w_{ti}/2} = \left(\prod_{i=1}^{N} \frac{p_{ti}}{p_{0i}} \right)^{(w_{0i}+w_{ti})/2} = P_{T}^{0,t}$$

This is the Törnqvist price index.

d. For the (arithmetic) Laspeyres price index:

$$P_{L}^{0,t} = \sum_{i=1}^{N} \frac{p_{ti}q_{0i}}{p_{0i}q_{0i}} \quad \text{and} \quad P_{L}^{t,0} = \sum_{i=1}^{N} \frac{p_{0i}q_{ti}}{p_{ti}q_{ti}}$$

Now, multiply these terms together to see whether they equal 1:

$$P_{L}^{0,t} \cdot P_{L}^{t,0} = \sum_{i=1}^{N} \frac{p_{ti}q_{0i}}{p_{0i}q_{0i}} \cdot \sum_{i=1}^{N} \frac{p_{0i}q_{ti}}{p_{ti}q_{ti}} \neq 1$$

For the Törnqvist price index:

$$P_{T}^{0,t} = \left(\prod_{i=1}^{N} \frac{p_{ti}}{p_{0i}} \right)^{(w_{0i}+w_{ti})/2} \quad \text{and} \quad P_{T}^{t,0} = \left(\prod_{i=1}^{N} \frac{p_{0i}}{p_{ti}} \right)^{(w_{ti}+w_{0i})/2}$$

Now, multiply these terms together to see whether they equal 1:

$$P_{T}^{0,t} \cdot P_{T}^{t,0} = \left(\prod_{i=1}^{N} \frac{p_{ti}}{p_{0i}} \right)^{(w_{0i}+w_{ti})/2} \cdot \left(\prod_{i=1}^{N} \frac{p_{0i}}{p_{ti}} \right)^{(w_{ti}+w_{0i})/2}$$

$$= \left(\prod_{i=1}^{N} \frac{p_{ti}}{p_{0i}} \cdot \frac{p_{0i}}{p_{ti}} \right)^{w_{0i}+w_{ti}} = \left(\prod_{i=1}^{N} 1 \right)^{w_{0i}+w_{ti}} = 1$$

J.2 a. $P_{L}^{0,1} = 1.0963$ $P_{P}^{0,1} = 1.0818$ (working to four decimal places)

b. $P_{GL}^{0,1} = 1.0913$ $P_{GP}^{0,1} = 1.0870$

c. $P_{T}^{0,1} = \sqrt{1.0913 \times 1.0870} = 1.0892$
   $P_{F}^{0,1} = \sqrt{1.0963 \times 1.0818} = 1.0890$

d. i. The percentage difference between the Fisher and the Törnqvist price index number is

$$\% \operatorname{diff}(F, T) = -0.01\%$$

ii. The percentage difference between the geometric Laspeyres and Passche index numbers is

$$\% \operatorname{diff}(GL, GP) = 0.39\%$$

iii. The percentage difference between the (arithmetic) Laspeyres and Paasche index numbers is

$$\% \operatorname{diff}(L, P) = 1.33\%$$

iv. The difference between the Törnqvist price index number and the Fisher price index number is very small; they are usually close in magnitude. The Laspeyres and Paasche price index numbers differ more than their geometric counterparts.

Although the difference between the (arithmetic) Laspeyres and Paasche price index numbers is greater than their geometric counterparts, the geometric mean of each pair is very similar.

Appendix E

Further reading

E.1 Manuals

The International Labor Organisation Consumer Price Index Manual: Theory and Practice is a very comprehensive guide to both index theory and practical price statistics relevant to the consumer price index. It consists of 23 chapters and a number of appendices. Some of the chapters on index theory are very technical, but many others are more practically focused. http://www.ilo.org/public/english/bureau/stat/guides /cpi/#manual

There is also a practical guide to producing consumer price indices, produced by the United Nations Statistics Division. http://unstats.un.org/unsd/iiss/Practical-Guide-to -Producing-Consumer-Price-Indices.ashx

For the UK Consumer Prices Index (CPI), there is a technical guide, which is updated every year; the current edition (at the time of writing this book) is the 2014 edition. http://www.ons.gov.uk/ons/rel/cpi/consumer-price-indices---technical -manual/2014/index.html

There is an International Labor Organisation manual for the Producer Price Index. The theoretical aspects are very similar to the corresponding chapters in the ILO CPI manual. http://www.imf.org/external/np/sta/tegppi/index.htm

E.2 Books

There are a number of index number books available. The simpler books were originally published in the 1960s and 1970s. There are more recent books, but they provide more advanced treatment of the subject. We list a selection of these books.

Index Numbers in Economic Theory and Practice by R.G.D Allen was originally published in 1975. A paperback edition was published in 2008 by Transaction Publishers.

A Practical Introduction to Index Numbers, First Edition. Jeff Ralph, Rob O'Neill and Joe Winton.
© 2015 John Wiley & Sons, Ltd. Published 2015 by John Wiley & Sons, Ltd.
Companion Website: http://www.wiley.com/go/ralph/index_numbers

Index Numbers by Walter Crowe was published in 1965 by MacDonald and Evans Ltd.

The Making of Tests for Index Numbers, by A. Vogt and J. Barta, was published by Springer-Verlag in 1997. It was published in honour of the 50th anniversary of the death of Irving Fisher. It presents a more modern update on the test approach that was championed by Fisher.

Two excellent, more advanced treatments of the subject are provided by the following books:

Index Theory and Price Statistics by Peter von der Lippe, Peter Lang, 2007.

Price and Quantity Index Numbers, Bert Balk, Cambridge, 2008.

A very recent addition to the set of index number books is an advanced treatment:

The Index Number Problem by Sydney Afriat, Oxford University Press, 2014.

E.3   Papers

The Ottawa Group is an international working group on price indices, which was formed in 1994 to bring together specialists to discuss research in the prices field. They meet every 2 years. The research papers from these meetings are available from their website, and they represent a great resource for serious students of the subject. http://www.ottawagroup.org/

Professor W. Erwin Diewert is a prolific author on index numbers and related topics; his page on the University of British Columbia website contains links to a large number of papers: http://www.economics.ubc.ca/faculty-and-staff/w-erwin-diewert/

Index

A Practical Introduction to Index Numbers, First Edition. Jeff Ralph, Rob O'Neill and Joe Winton.
© 2015 John Wiley & Sons, Ltd. Published 2015 by John Wiley & Sons, Ltd.
Companion Website: http://www.wiley.com/go/ralph/index_numbers